Risk, Systems and Decisions

Health, environment, security, energy, technology are problem areas where man-made and natural systems face increasing demands, giving rise to concerns which touch on a range of firms, industries and government agencies. Although a body of powerful background theory about risk, decision, and systems has been generated over the last several decades, the exploitation of this theory in the service of tackling these systemic problems presents a substantial intellectual challenge. This book series includes works dealing with integrated design and solutions for social, technological, environmental, and economic goals. It features research and innovation in cross-disciplinary and transdisciplinary methods of decision analysis, systems analysis, risk assessment, risk management, risk communication, policy analysis, economic analysis, engineering, and the social sciences. The series explores topics at the intersection of the professional and scholarly communities of risk analysis, systems engineering, and decision analysis. Contributions include methodological developments that are well-suited to application for decision makers and managers.

Les Servi

Mission Dependency Network Analysis

Mathematical Foundations for Modeling and Practice

Springer

Les Servi
Lincoln, MA, USA

ISSN 2626-6717 ISSN 2626-6725 (electronic)
Risk, Systems and Decisions
ISBN 978-3-032-11259-0 ISBN 978-3-032-11260-6 (eBook)
https://doi.org/10.1007/978-3-032-11260-6

This Springer imprint is published by the registered company Springer Nature Switzerland AG
The registered company address is: Gewerbestrasse 11, 6330 Cham, Switzerland

To my wife, Susan, whose steady support and patience helped make this book possible and to Amelia, Eli, Franklin, Ila, and Joseph who may grow up with entirely different interests, but who, should they one day stumble upon these pages, may discover a few traces of themselves in this endeavor.

Foreword

"If you know the enemy and know yourself, you need not fear the result of a hundred battles. If you know yourself but not the enemy, for every victory gained you will also suffer a defeat. If you know neither the enemy nor yourself, you will succumb in every battle." Sun Tzu, *The Art of War*

"You just got hosed." Lt John Boyd, Boyd—*The Fighter Pilot Who Changed the Art of War* (page 86)

Making good decisions faster than your adversary/opponent greatly increases your probability of winning. Sun Tzu essentially stated this around 500 BC. John Boyd made this modern in his famous Observe Orient Decide Act (OODA) loop development based on his "Aerial Attack Study" and follow-on work on airplane dogfights. Today with live data feeds, digital twins of systems of systems, and generative adversarial nets that leverage cutting edge artificial intelligence, gaining decision advantage requires better/faster operations analysis than that of your adversary/opponent/pacing challenge.

Building out, operating, and assessing this capacity by operations analysts takes significant funding; funding that might otherwise go to munitions, spare parts, cyber defense, training hours, or fuel just to name five potential opportunity cost choices.

1. How should one assess such zero sum tradeoff decisions?
2. What confidence do you have that those decisions are correct?
3. How sensitive are those decisions to your input variables?
4. How sensitive are those decisions to your adversaries reactions/moves/ countermoves?

The book, *Mission Dependency Network Analysis*, written while Dr. Les Servi served as MITRE Chief Scientist for Cyber Operations Research, provides equations that may be able to provide better answers faster than can be achieved today.

To frame Dr. Servi's book, consider two adversaries, perhaps chess players, checkers players, boxers, pilots, or Joint Task Force Commanders preparing to compete. For simplicity, call these adversaries Player Red and Player Blue. Both Player Red and Player Blue will seek to achieve these goals simultaneously:

1. Execute a winning strategy.
2. Blunt the adversary's winning strategy.
3. Seize and maintain the initiative to force the adversary to react vice to act.
4. Be able to fund 1–3 above until the adversary dies / resigns / loses / bankrupts / surrenders.

With this framing in mind, consider leveraging Dr. Servi's book and equations to solve the competition's system-of-systems solutions to win and blunt faster and cheaper.

Using the boxing example, say Player Red believes that stamina and speed are their comparative advantages so they choose to avoid closing until the later rounds off the fight. Player Blue believes they can knock out Player Red because their comparative advantages are strength and reach so they seek to close with the opponent and land heavy blows early and often, before they lose stamina because they punched themselves out. Let's say they both trained equally as a proxy for how well they are funded to keep this example simple. Now the one that can force the tempo they want is more likely to seize the initiative and win.

Now scale up that example to cyber warfare for a quick excursion. Say Player Red and Player Blue are equally funded. Both players must choose how much to fund:

1. Gathering of cyber intel
2. Manning, training, and equipping for cyber offense
3. Manning, training, and equipping for cyber defense
4. Funding assessments of decision advantage

Say Player Red put 25% of available funds to each. Player Blue put 40% to Intel and cyber offense and only 10% to defense and decision advantage. Who would win? How sure are you? What if Player Blue played the exact opposite strategy, how would that change things? How sensitive are the results to assumptions that are swags (scientific wild ass guesses), vice a known and tested digital twin of the system?

The smart analyst will both leverage Dr. Servi's equations and improve upon them and publish those results. This is why Dr. Servi wrote this book.

In closing, in Kurtis Minder's 2025 book *Cyber Recon: My Life in Cyber Espionage and Ransomware Negotiation*, Kurtis emphasizes that the United States is currently fighting World War III in cyber space and the United States is losing. Perhaps this will spur analysts and decision makers to leverage Dr. Servi's book to ask better questions, fund better strategies to win, blunt and seize the initiative cheaper, and understand how defendable those strategies are to Congress and against the adversaries of the United States.

Captain Robert Chandler Swallow,[1] USN (Ret.)

[1] The views expressed in this foreword are the author's own and do not reflect the official policy or position of the Department of the Air Force, Department of the Navy, Department of Defense, or the US Government. The author received no financial support for the research, authorship, and/or publication of this article.

Preface

> *"Now the general who wins a battle makes many calculations in his temple before the battle is fought. The general who loses a battle makes but few calculations."*
>
> —Sun Tzu (c. 500 B.C.) (Tzu 1910, 1963)

Mission dependency network analysis models systems of systems that are increasingly complex, interconnected, critically important, and inherently vulnerable. They may represent systems as vast as global logistics networks or as contained as embedded vehicle control systems. Alternatively, they can capture purely functional constructs such as coordinated operational tasks. In all cases, they are susceptible to failure from accident, malicious action, natural disaster, or internal degradation.

Operations Research (OR) provides powerful tools to enhance network design and identify vulnerabilities for defensive or offensive purposes. Despite its widespread use in practice, OR remains insufficiently appreciated outside specialized circles. For nearly a century, it has advanced the design, performance, and effectiveness of operational tasks, achieving, for example, pivotal impact during World War II (Burchard 1948). In the years since, OR has extended its reach across a wide spectrum of applications in industry and government.

Concurrently, a quiet revolution has expanded the reach of OR, enabling analyses once considered out of reach. From 1988 to 2004, solution times for large linear programs fell by more than five million-fold, driven by roughly 1,600-fold improvements in hardware and about 3,300-fold advances in algorithms (Bixby 2002).

In 2022, the Military Operations Research Society (MORS) selected "Analytics as a Force Multiplier" as its annual theme, underscoring these accelerating advances in OR as well as the emerging promise of Large Language Models (Servi 2022). The intent was to reframe the view of analytics as a tool, a modern day *bicycle for the mind* (Jobs 2006)—one that amplifies the power, agility, and impact of the systems we already possess. This book may be viewed as one answer to that call.

To bring OR and mission dependency network analysis together, one must:

- *Map the mission dependency network*: Model the components required for mission success as a dependency graph. In this representation, network nodes correspond to systems or functions, and edges capture the dependencies between them. For clarity throughout this book, treat systems as physical nodes and functions as functional nodes. Where both are present, identify the physical nodes that enable or support specific functional nodes, and, when possible, trace the physical node pathways to functional nodes whose disruption would degrade mission success.
- *Quantify dependencies*: Develop a parameterized framework that quantifies how dependencies between mission components are represented (e.g., as weights, logical rules, or other functions), and how effects propagate through the mission dependency network to impact mission success.
- *Estimate dependency parameters*: When dependency parameters are unknown, due to limited access to knowledgeable experts, use statistical inference methods to estimate them if real or simulated data to do so is available. When expert judgment is available, use statistical inference to complement or validate that judgment.
- *Assess criticality*: Develop methods to identify critical nodes within a mission network shaped by operational and informational constraints and using a definition of criticality grounded in the mission's needs.
- *Hedge against known unknowns*: Account explicitly for inaccuracies in assumed dependency parameters and then perform analysis that reflects the decision maker's tolerance for risk and preferred level of safeguards against data uncertainty. Use intervals rather than point estimates whenever doing so better captures the degree of uncertainty.
- *Model time*: Model disruption dynamics by explicitly parameterizing detection, response, and recovery durations. When feasible, derive resilience metrics. Select time resolution appropriate to the use case, ranging from digital twin level detail to aggregated time steps that summarize underlying micro-dynamics.

This book explores these considerations and establishes a vocabulary and analytical framework to address them. It presents a body of analysis developed internally at The MITRE Corporation while the author was Chief Scientist for Cyber Operations Research. While focused on practical applications, this book also, where needed, advances the state of the art by overcoming obstacles missing in the current academic literature. Some chapters have highly technical parts, which may require a deep understanding of Operations Research to fully grasp them. Nonetheless, the book is designed to engage three distinct audiences:

- *Senior decision makers* seeking a strategic, high-level understanding and insight into the art of the possible. While likely to pass over the detailed Operations Research mathematics, this reader may stretch their goals and perspectives and then demand, using a more precise language, solutions that leverage more modern methods and more nuanced goals. This reader should focus on chapter introductions, figures, and examples to learn what is possible.

- *Operational practitioners* seeking to apply the algorithms or problem formulations, with limited interest in the underlying mathematical rigor or theoretical details. This reader should focus on the algorithms presented, the linear programs formulated, and the illustrative examples with an eye toward adapting approaches in this book to their practical applications.
- *Applied researchers* in academia, industry, or government interested in refining, extending, or adapting the methods to systems relevant to missions or domains of interest to them. Alternatively, this reader may have minimal interest in the domain of this book, but wishes to use the book as a means of expanding their knowledge of Operations Research by example rather than from a more classical Operations Research textbook. This reader should focus on the detailed development and proofs of the various problem formulations.

There is a concern that rigorous analysis is infeasible because even basic inter-system dependency data may be unavailable. Chapters 7 and 8 start to address this by explicitly modeling imprecise inputs. Chapter 2 shows how to use experiments or simulations, when available, to statistically infer the necessary parameters. In many situations, the practical course is to proceed with the best available estimates and run an iterative, decision focused process to surface vulnerabilities, bound mission effectiveness, and target the next, most decision-relevant data to gather. Even with incomplete data, this process may reveal concrete, high-leverage actions and pinpoints where additional evidence will most improve the decision without relying on the hope that purely qualitative or a nearly qualitative pass will suffice.

Key takeaways of this book are listed in Table 1, along with a mapping to the chapters that mathematically address each.

If this book succeeds, it will not be the last word on these topics, but a beginning meant to inspire (i) senior leaders to unlock new levers of impact, (ii) practitioners to sharpen and extend the algorithms for real-world use, and (iii) researchers to push the theory in directions both rigorous and relevant—far beyond what this book imagines. This would be facilitated by multiple forums: some brief formats for a broad audience, some targeted medium to long ones to help translate ideas in this book in a customized manner for each audience, and, most importantly, many shared forums where all three audiences are in the same room. When one plans together, the math changes: $1 + 1 + 1 \gg 3$.

Lincoln, MA, USA Les Servi

Table 1 Key takeaways

		Chapters								
Topic	Description	1	2	3	4	5	6	7	8	9
1. Adapt to mission type	Use either binary missions or effectiveness mission types according to what is most fitting to the mission of interest.									
– 1A. Binary missions	A mission will either succeed or fail.			X	X	X				
– 1B. Mission effectiveness	A mission operability effectiveness will be scored in [0 100].	X	X	X			X	X	X	X
2. Define and analyze criticality according to the decision maker's preferred definition of criticality	Define "criticality" carefully as it encodes the decision maker's goals and drives the computed recommended decision. When resources cannot disable the mission outright, this definition impacts how components are ranked. We consider four definitions.									
– Criticality definition option 2A	Rank components by the expected increase in the probability of mission failure when each is disabled individually, averaged over uncertainty in all other components.					X				
– Criticality definition option 2B	Identify the subset of components whose joint disruption causes mission failure (cutset analysis).					X				
– Criticality definition option 2C	Select the set of components whose degradation maximizes diminishment of the mission effectiveness subject to resource constraints.		X		X		X	X	X	X
– Criticality definition option 2D	Measure the marginal change in mission effectiveness from the marginal degrading of a single component (sensitivity analysis).	X		X						
3. Anticipate the adversary's responses when making decisions	Find one's best action only after anticipating one's adversary's response. Antifragility might be found, i.e., conditions under which disruption can ultimately strengthen the targeted system, e.g., a disruption makes retiring legacy and modernizing architecture economically inevitable ultimately leaving the system stronger than its pre-attack baseline.				X				X	X

(continued)

Table 1 (continued)

		Chapters								
Topic	Description	1	2	3	4	5	6	7	8	9
4. Incorporate disrupt vs. monitor tradeoff	Balance degrading an adversary's capability against the intelligence value of continued access and monitoring (intelligence gain / loss analysis).						X			
5. Deal with uncertainty in data	Model imperfect dependency parameter input data with intervals (not just points) and optimize decisions which tradeoff mission impact and decision maker's risk tolerance for data uncertainty. Neither plan for a data accuracy worst case nor best case; be more nuanced.		X		X			X	X	
6. Acknowledge relationship between physical systems and functional capability within a mission	Account for the operability of combinations of physical systems that support the functional capabilities required for mission success as well as physical systems needed to transverse before one can impact functional capability.				X	X	X			X
7. Time matters	When appropriate, capture detection delays, response actions, and recovery durations to quantify temporal resilience.									X
8. Focus on effects, not modalities	Avoid needlessly differentiating kinetic from non-kinetic actions. It is the effect that matters, not the modality to the mathematics in this book. Although not explicitly discussed, the fact that this book does not need to differentiate among modality types suggests that such siloing may distract from good analysis.									

References

Bixby RE (2002) Solving real-world linear programs: a decade and more of progress. Oper Res 50(1): 3–15

Burchard JE (1948) Q.E.D.: M.I.T. in World War II. Wiley, New York, pp 86–102

Garvey PR, Pinto CA (2012) Advanced risk analysis in engineering enterprise systems. Chapman-Hall/CRC-Press, Taylor & Francis Group (UK), Boca Raton. ISBN 9781439826140

Jobs S (2006) Computers are like a bicycle for our minds. YouTube. https://www.youtube.com/watch?v=ob_GX50Za6c

Servi LD (2022) Analytics as a force multiplier. Phalanx. Fall Issue 55(3): 2–4

Servi LD, Garvey PR (2017) Deriving global criticality conditions from local dependencies using functional dependency network analysis (FDNA). Syst Eng 20(4):297–306
Tzu S (1910) The art of war: the oldest military treatise in the world (L Giles, Trans.) Project Gutenberg. https://www.gutenberg.org/ebooks/132. Chapter 1 (26)
Tzu S (1963) The art of war (SB Griffith, Trans.) Oxford University, Oxford. ISBN 978-0-19-501540-9. Chapter 1 (26)

Acknowledgments This book would not be possible without the support of The MITRE Corporation, and in particular, the Cyber Investment Area Leads of years past, Dr. Vipin Swarup and Dr. George Roelke, to whom I am deeply grateful. The MITRE Corporation is one of the few organizations which still has an Operations Research department. I wish to acknowledge and thank its leader, Dr. John Stine, for his superb support and guidance of this department as well as the current and past leaders of the Modeling and Simulation Innovation Center, which houses the Operations Research Department, Dr. Phil Barry, Dr. John Pang, Dr. Chris Glazner, and Jennifer Tress. They all created the environment conducive to great work, and which attracted an extraordinarily talented staff to work with. I am also grateful to R. Chandler Swallow, Director, Program Assessment and Evaluation (SAF/AE), Office of the Secretary of the Air Force, for generously making time to discuss this book and for contributing the Foreword.

Numerous MITRE staff who were not directly involved in any specific chapter but with whom I have had many technical discussions that might have found their way into this book include Dr. Doug Altner, Dr. Paul Garvey, Dr. Hannah Horner, Dr. Berk Öztürk, and Dr. She'ifa Punla-Green. Jhanny Jimenez Feliz, while a student at Notre Dame Cristo Rey High School (A school that works) and an intern at MITRE, provided invaluable software support for Chap. 1, before becoming an undergraduate at the Massachusetts Institute of Technology.

Acknowledgments would not be complete without naming my thesis advisor, Professor Y. C. Ho (formerly of Harvard University), Professor Julian Keilson (formerly of University of Rochester and MIT), and Dr. Daryl Daley (formerly of Australian National University) who were each early collaborators with me of numerous papers and who, by example, taught me how to do applied research.

Finally, I wish to express my deep gratitude to Dr. Edward Hua, who provided meticulous and thoughtful peer review of each chapter. His efforts materially improved many sections and were invaluable to the final product.

All errors and omission in this books are, however, the sole responsibility of mine alone.

Competing Interests The author has no competing interests to declare that are relevant to the content of this manuscript.

Reference

A school that works. YouTube. https://www.ndcrhs.org/apps/video/watch.jsp?v=100118

Permissions

The author's affiliation with The MITRE Corporation (MITRE) is provided for identification purposes only and is not intended to convey or imply MITRE's concurrence with, or support for, the positions, opinions, or viewpoints expressed by the author.

Chapter Overview

You've got to have multiple models – because if you just have one or two that you're using, the nature of human psychology is such that you'll torture reality so that it fits your models.

Charlie Munger (1994)

Part I: Introduction to Criticality Analysis of mission dependency networks with Functional Components

Part I introduces core concepts and definitions for mission dependency network analysis with a focus on identifying centers of gravity in systems of systems. Part I assumes a continuous mission effectiveness score in [0,100]. Later parts of this book relax this assumption to a binary success/failure score. Part I first uses a specific representation to quantify dependencies between network components and then proposes two more general frameworks: one based on continuous variables and one based on discrete variables. Together, these chapters provide the foundation for the remainder of the book.

Chapter 1: Assessing Node Criticality from Local Dependencies

This chapter tackles a core challenge in mission dependency network analysis: identifying, in a computationally efficient manner, which parts of a network are most critical to mission success. It introduces both the Functional Dependency Network Analysis (FDNA) representation [3] for quantifying how components depend on one another, and a simple yet powerful metric of node criticality [8]. It demonstrates, for the first time, that both node operability and this notion of criticality can be obtained as solutions to linear programs (LPs). This enables the use of modern open source and commercial LP solvers at scale to compute them. Key ideas are illustrated using variations of a simple four node network, while noting that the approach is scalable to networks with hundreds or even thousands of nodes. The algorithm maintains near linear time efficiency for realistic mission network topologies.[1]

Chapter 2: A Continuous Universal Framework for mission dependency networks

To promote mission effectiveness and criticality analysis algorithm reuse across heterogeneous models now and in the future, this chapter standardizes a *continuous* parameter framework to quantify dependencies between mission network components. The framework subsumes common dependency models as special cases. This chapter then examines this framework in the context of finding the best actions for a defender or an attacker under two scenarios: (i) the dependency parameters are known precisely and (ii) the dependency parameters values are only known to be in an interval.

Chapter 3: A Discrete Universal Framework for mission dependency networks Applied to System Identification and Critical Node Analysis

Complementing Chap. 2, this chapter standardizes a *discrete* parameter framework to quantify dependencies between mission network components as they are related to mission effectiveness. Additionally, this chapter presents an identification algorithm that estimates the value of the parameters for this general framework using input from simulation or experimental data. This could be done either in the absence of or to complement the more time-consuming efforts of subject matter experts (SMEs). A second algorithm is introduced which uses these parameters to find the mission components that most significantly influence the mission effectiveness. This chapter concludes with a summary of the empirical performance of both algorithms.

Part II: Mission Dependency Network Analysis with Physical and Functional Components

Part II builds on Part I by adding physical nodes, nodes which correspond to systems, and their relationship to functional nodes which correspond to mission functions or tasks. It addresses two practical realities: (i) functional nodes are

[1] Although this chapter uses FDNA, the approach extends beyond FDNA.

enabled or supported by combinations of physical nodes, and (ii) an adversary might need to traverse a network of physical nodes before reaching a functional node which can be degraded. Chapters 4 and 5 take two complementary approaches to the first issue. Chapter 6 focuses on the second. Chapters 4 and 5 assume the mission has a binary success/failure score whereas Chap. 6 assumes the mission has an effectiveness score in [0,100].

Chapter 4: Disrupting Functional Pathways Having Supporting Physical Systems: An OR Approach

In contrast to previous chapters, this chapter employs an Operations Research formulation that represents a mission success as achieved by successful completion of a sequence of functional tasks. Each task (functional node) is successful only if its required combination of physical systems (physical nodes) is operable. The mission is successful only when there exists a start-to-end path consisting solely of operable tasks at the time required. Within this framework, this chapter introduces a method to identify the most critical physical nodes in the mission network. The method is then extended to handle uncertainties in physical node attributes, such as the cost of disabling them and their individual impact. Numerical experiments on networks with up to ten thousand nodes demonstrate that this method is computationally efficient, exhibiting near linear runtime efficiency.

Chapter 5: Disrupting Functional Pathways Having Supporting Physical Systems: A BSI Approach

Tractable criticality analysis of large-scale systems is essential. One approach that supports this goal is the use of the Birnbaum Structural Importance (BSI) measure. BSI is a long-standing, well-established metric for assessing nodal criticality in reliability studies. While BSI has traditionally been applied to purely physical networks, this chapter extends the BSI approach to hybrid networks that incorporate both physical and functional nodes. Here the physical nodes which support the functional nodes are included in the analysis. An exact BSI computation scales exponentially with the number of nodes, requiring $O(2^N)$ operations for a network with N nodes. To overcome this practical limitation, this chapter introduces an approximation algorithm with complexity $O(\epsilon^2 N^2)$, where ϵ represents the desired error bound. This approximation algorithm completes in about 30 seconds for a 112 node network, and is many orders of magnitude faster than the computing requirements for finding the exact BSI value.

Additionally, this chapter presents (i) a new, fast method for the actionable need to identify the top m nodes with the highest BSI scores; (ii) a new, fast method for computing cutset nodes, whose joint disabling guarantees mission failure; and (iii) a new measure of network resilience derived from the BSI perspective. Examples illustrate usage and report computational timings.

Chapter 6: Traversing a Network of Physical Nodes to Attack Functional Components of a mission dependency network

For some missions, attacks begin by traversing a network of physical nodes with the aim of compromising high value functional nodes which determine the mission's effectiveness. To address the need to analyze such missions, this chapter builds on the purely functional mission dependency networks introduced in Part I, but adopts a different modeling approach than the previous two chapters. Specifically, it introduces a physical network for which each node has a known traversal cost. Certain endpoints in this physical network connect to functional nodes. To disrupt the mission network, an attacker must enter through a physical node, navigate the physical network node by node to reach, and compromise one of the functional nodes. The resulting mission effectiveness impact, measured between 0 and 100, is computed from the functional dependency network equations encoded in the mission dependency network model.

This chapter provides the analytical foundation for identifying optimal attack plans under four progressively realistic scenarios. It concludes with a simple, illustrated example that highlights how these scenarios differ in structure and outcome. The most realistic scenario captures intelligence gain/loss analysis, i.e., finding the tradeoffs between disrupting the mission and having continued access for monitoring the adversary.

Part III: Mission Dependency Network Analysis with Incomplete System-of-System Knowledge and Temporal Considerations

Part III addresses two often overlooked realities in mission dependency network analysis: (i) dependency data are imperfect, and (ii) disruptions are time dependent, as detection, response, and recovery are not instantaneous. Chapters 7 and 8 treat parameter uncertainty by modeling dependencies with interval values rather than point estimates, helping decision makers balance mission objectives against risks arising from model uncertainty. Chapter 9 introduces temporal dynamics, capturing detection delays, response, and recovery to quantify mission resilience over time.

Chapter 7: Robust Mission Network Analysis: The Theory

Real-world data are costly to obtain and seldom precise: We rarely know exactly how strongly one component depends on another or the exact costs of alternative actions. Building on the robust optimization literature, this chapter examines linear programs where the coefficients of the objective function and constraints are all known to lie within specified intervals, but not known exactly. This chapter shows how to strike a balance between maximizing the objective and effectively managing uncertainties without overplanning for worst cases that may never materialize. While the focus of this chapter is on the theoretical analysis of linear programs, this chapter shows how to translate the results into mission network defense decisions when dependency parameters are known unknowns.

Chapter 8: Robust Mission Network Analysis: The Practice

Building on the robust optimization methods from Chapter 7, this chapter finds the best defensive courses of action (COAs) under varying assumptions about the accuracy of dependency parameters and compares the resulting optimal COAs empirically. It then introduces a game theoretic attacker defender model to show how anticipating the opponent's response alters COA choices. Finally, it explores antifragility in network defense—cases where disruption can paradoxically ultimately strengthen the defender's mission.

Chapter 9: Incorporating Time into mission dependency network Analysis

For some missions, the role of time is essential. This chapter extends the book's primary focus on static analyses to now include temporal dynamics. Specifically, it examines defensive or offensive decisions that create impacts that propagate through the mission network in time. The chapter explores the implications of these dynamic interactions on the mission network's resilience and provides some illustrative examples of the presented approach.

Reference

Munger C (1994) A Lesson on Elementary, Worldly Wisdom As It Relates to Investment Management & Business. Speech, USC Business School, Transcript. https://www.safalniveshak.com/wp-content/uploads/2012/08/Lesson-on-Elementary-Worldly-Wisdom-Charlie-Munger.pdf:1

Contents

Contributors

Earlier drafts of a number of chapters of this book were either co-written with MITRE colleagues or heavily influenced by previous discussions as indicated below.

- Chapter 2: Damon Frezza and Kael Stilp
- Chapters 3, 7, 8: Damon Frezza and Erica Mason (formerly of MITRE, now at Spotify)
- Chapters 4, 5, 9: Anthony Rojas (formerly of MITRE, now at Syngenta)
- Chapter 6: Kael Stilp

List of Figures

List of Tables

Part I
Introduction to Criticality Analysis of Mission Dependency Networks with Functional Components

Chapter 1
Assessing Node Criticality from Local Dependencies

For want of a nail the shoe is lost;
for want of a shoe the horse is lost;
for want of a horse the rider is lost;
for want of a rider the message was lost;
for want of a message the battle was lost;
then the war was lost.

Benjamin Franklin (1795)

1.1 Introduction

1.1.1 Overview of Problem

The imperative to affordably embed resilience into critical networks continues to grow as systems of systems become intertwined globally. Achieving this demands an understanding of the components, interdependencies, and vulnerabilities within these systems of systems. Network graphs provide a powerful framework for modeling such systems to enable the development of algorithms that pinpoint critical nodes. One notable approach to capturing these interdependencies, Functional Dependency Network Analysis (FDNA) (Garvey and Pinto 2009, 2012; Servi and Garvey 2017), uses real-valued mathematical functions. FDNA enables the modeling and quantification of operational effectiveness at both the node and network levels by capturing the inflows and outflows of operability across the systems of systems.[1]

While this entire quotation has been attributed to Benjamin Franklin, it is believed that the last three lines are a misattribution and, in fact, came from variations of the words of James Baldwin (1912), Fifty Famous People. The Horseshoe Nails. American Book Company: 51–54.

[1] While this book emphasizes FDNA due to its wide use, the concepts and methodologies presented are broadly applicable and can be generalized to alternative dependency frameworks.

L. Servi, *Mission Dependency Network Analysis*, Risk, Systems and Decisions,
https://doi.org/10.1007/978-3-032-11260-6_1

Ensuring the resilience of critical systems in challenging environments requires the ability to determine which elements or assets have the most significant impact on the mission. This chapter reexamines the recently introduced metric, the Degradation Index[2] (Servi and Garvey 2017), which measures how a small change in one node's operability value impacts the overall mission's operability value. The Degradation Index can be viewed as transforming local information about a functional network to global information about the network's operability and critical components and can be considered an example of sensitivity analysis. The first algorithm to compute this Degradation Index in this setting (Servi and Garvey 2017) used an iterative algorithm that sequentially examined all the nodes and their dependencies. This chapter will describe an alternative approach to compute the Degradation Index by solving a set of linear programs. This approach leverages the relentless increase in speed developed for solving linear programs. For example, between 1989 and 2003, the Gurobi solution speed increased by over five orders of magnitude (Bixby 2015).

In the following sections of this chapter, the basic FDNA dependency representation is mathematically described in (1.1) along with a review of a previous approach to both quantify the operability and estimate the criticality of each node based on the analysis in Servi and Garvey (2017). Next, a new method to do both is introduced using a sequence of six linear programs that can leverage the speed of commercial linear programming optimization packages. Finally, to illustrate key concepts, three variations of a simple example are presented, as well as a sequence of larger examples. The examples illustrate potential hurdles and mitigations of this linear programming approach, as well as the computational speed of this approach.

Criticality analysis is our entry point into the topic of this book, Mission Dependency Network Analysis. In this chapter we build a baseline model under four simplifying assumptions of appropriate for the mission of interest: (1) Mission effectiveness is scored on a 0–100 scale, (2) criticality is defined and measured by the Degradation Index (DI), (3) all dependency data are known exactly, (4) attackers and defenders act myopically, i.e., neither anticipates the other's reactions, and (5) time dynamics are not of pivotal importance. These simplifications provide a clear and practically useful starting point. Subsequent chapters relax them.

1.1.2 Technical Background

This section introduces a set of equations, the FDNA equations, which describe a useful framework for capturing the dependencies between components of a mission network. In addition, this section introduces a recently proposed measure of

[2] The term 'Degradation Index' as used in this chapter refers to a concept distinct from the one employed by the U.S. Navy, which uses the same term with a different meaning (Commander, Navy Region Hawaii 2023).

criticality of nodes, the Degradation Index, and a previous approach for computing it.

Consider a mission represented by a dependency network where P_i is the operability of node i, $C(i)$ is the set of nodes whose operability impacts node i, and $\mathcal{L}$ is the set of leaf nodes that, by definition, do not depend on any other nodes, i.e., if $i \in \mathcal{L}$, then $C(i)$ is the empty set. Also, define P_i^0 as the given operability of leaf node i.

For illustrative purposes, consider a system consisting of a single satellite receiver, two relay stations that connect to this satellite receiver, Relay Station A and Relay Station B, and a single ground station connected to the two relay stations. To analyze the criticality of the ground station on the mission, the full operability of the satellite receiver, assumptions must be made about the dependency between (1) the satellite and Relay Station A, (2) the satellite and Relay Station B, (3) Relay Station A and the ground station, and (4) Relay Station B and the ground station. Assumptions will be made of these four dependencies in a series of examples in Sect. 1.3 based on Eq. (1.1) and summarized in Table 1.1. For this example there is only a single leaf node, the ground station, and the four rows of Table 1.1 list in the fourth column the nodes that are dependent on the nodes in the second column.

(P0): The FDNA equations are

$$P_i = \begin{cases} \min\left[\min_{j\in C(i)}\left[P_j+\beta_{ji}\right], \sum_{j\in C(i)}\frac{\alpha_{ji}}{|C(i)|}P_j+100\left(1-\sum_{j\in C(i)}\frac{\alpha_{ji}}{|C(i)|}\right)\right], & \forall i\notin\mathcal{L}, \\ P_i^0, & \forall i\in\mathcal{L}. \end{cases} \tag{1.1}$$

Equation (1.1) has a unique solution, which can be solved recursively or sequentially, starting with the leaf nodes (Servi and Garvey 2017).

One might interpret the variable α_{ji} as capturing the relative strength of dependencies of node i's dependent nodes, referred to as child nodes of node i throughout this book and inversely, and node i is referred to as the parent node of its child nodes. The variable β_{ji} captures the criticality of each dependent node's operability and the extent it can, by itself, reduce the parent's operability. The α_{ji} terms are referred to as the *Strength of Dependency (SOD)* term, and the β_{ji} terms are referred to as the *Criticality of Dependency (COD)* terms (Garvey and Pinto 2009). Equation (1.1) forces the stronger influence between these two types of dependency parameters to prevail.

Table 1.1 An illustrative simple four-node example

Node ID, j	Node name	Dependent node ID, i	Dependent node name	α_{ji}	β_{ji}
1	Satellite Receiver	2	Relay Station A	0.80	10
1	Satellite Receiver	3	Relay Station B	0.80	10
2	Relay Station A	4	Ground Station	0.80	100
3	Relay Station B	4	Ground Station	0.75	15

We note, parenthetically, a less used variation of FDNA is the case of FDNA with logical OR nodes (Servi and Garvey 2017), which replaces (1.1) for some or all of the nodes:

$$P_i = \begin{cases} \max_{j \in C(i)} \left[\min \left[P_j + \beta_{ji}, \alpha_{ji} P_j + 100(1 - \alpha_{ji}) \right] \right], & \forall i \notin \mathcal{L}, \\ P_i^0, & \forall i \in \mathcal{L}. \end{cases} \tag{1.2}$$

The concept of logical OR nodes is used to convey that each individual child node, by itself, might impact its parent node independent of the other children. This chapter will focus exclusively on the classical form (1.1) rather than (1.2) or a hybrid of (1.1) and (1.2). However, the methods described in these chapters are applicable to all forms of FDNA.

The concept of a node's criticality can be defined in various ways, making it impossible to compare the relative effectiveness of each approach. To effectively engineer systems, it is essential to have a broad range of critical measures. This chapter explores one such measure, the Degradation Index, $\mathcal{D}_i = \frac{dP_1}{dP_i}$, which is highly efficient to compute numerically, and empirically very useful despite its simplicity. This criticality metric was first examined (Servi and Garvey 2017). It can be interpreted as the change of the network mission objective as a result of a small change in the operability of a single node. They computed it in two steps:

STEP 1: Sequentially solve for P_i, and $D(e) = \frac{\partial P_i}{\partial P_j}$: $\frac{\partial P_i}{\partial P_j}$ is the marginal impact of the operability across edge $e = (i, j)$ of the dependency network, i.e., the marginal impact of the operability of node j directly on the operability of node i. Depending on which component of (1.1) dominates the minimization, $D(e)$ equals 1 or α_{ji}. This step requires carefully selecting an order to perform the calculations to ensure that all information needed is known in time to perform the calculation.

STEP 2: Sequentially identify $\mathcal{P}(1, i)$, the set of all paths in the dependency network from node 1 to node i, and observe that

$$\mathcal{D}_i = \frac{dP_1}{dP_i} = \sum_{\{p:p \in \mathcal{P}(1,i)\}} \left(\prod_{\{e:e \in \mathcal{E}(p)\}} D(e) \right), \tag{1.3}$$

where $\mathcal{E}(p)$ is the set of edges in path p.

Equation (1.3) can be interpreted as follows: The product, using the chain rule, corresponds to the change of the mission node's, i.e., node 1's, objective value with respect to a change in node i's objective along path p, as a result of the cascading impact from the change along each edge of path p. The summation adds all these impacts together for all of the different paths from node 1 to node i.

The new approach presented in this chapter requires a nearly linear computation time for typical mission networks, and it is simpler to execute and is expected to have faster execution times for similar mission networks.

1.2 Technical Approach

This section introduces a new approach for solving the FDNA equation (1.1), using six related linear optimization problems. These problems can be solved very fast using a standard LP solver and, typically as a byproduct, permit one to quickly compute the criticality of each of the nodes.

The six related problems are:

(P1): a linear program that always has an optimal solution which solves the FDNA equation (1.1) but typically has other optimal solutions which do not

(P2): a variation of (P1) with a single solution, which is guaranteed to solve the FDNA equations

(P3): the dual of the linear program of (P1), whose solution can be used to quantify the criticality of each node, the Degradation Index $\mathcal{D}_i$, for the case of (P3) having a unique solution

(P4): a variation of (P3), which determines whether or not (P3) has a unique solution

(P5): a variation of (P3), which finds the lower bound of the Degradation Index, $\mathcal{D}_i$, of node i for the case of (P3) not being unique

(P6): a variation of (P3), which finds the upper bound of the Degradation Index, $\mathcal{D}_i$, of node i for the case of (P3) not being unique

This approach is summarized in Fig. 1.1.

Each problem is described in detail and then put in a tight matrix form conducive toward being solved using a commercial linear program.

By convention, node 1 denotes the mission operability node and is the only node with the property that no node depends on it, i.e., $1 \notin C(i)$ for all nodes i. Also, all nodes, i, which do not depend on any other nodes, i.e., $C(i)$ is an empty set, are defined as *leaf nodes*.

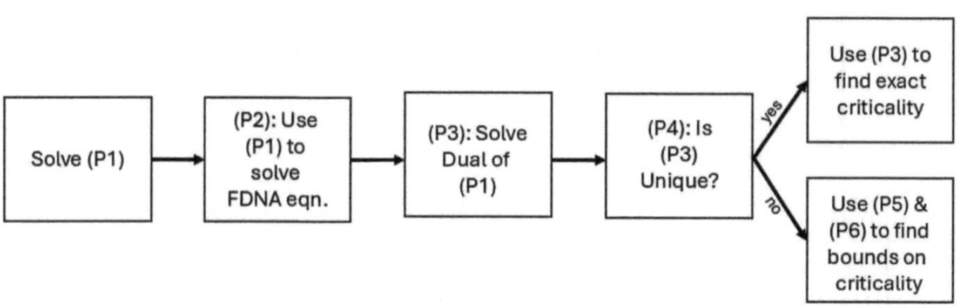

Fig. 1.1 Computational flowchart to finding criticality metrics

1.2.1 The Linear Problem Used to Solve the FDNA Equations

The first linear program takes the FDNA equation (1.1) and rewrites it as a related optimization problem.
(P1):

$$\max_{P} P_1 \tag{1.4}$$

subject to

$$P_i \le P_j + \beta_{ji}, \qquad \forall j \in C(i), \forall i \notin \mathcal{L}, \tag{1.5}$$

$$P_i \le \sum_{j\in C(i)} \frac{\alpha_{ji}}{|C(i)|} P_j + 100\left(1 - \sum_{j\in C(i)} \frac{\alpha_{ji}}{|C(i)|}\right), \qquad \forall i \notin \mathcal{L}, \tag{1.6}$$

$$P_i \le P_i^0, \qquad \forall i \in \mathcal{L}, \tag{1.7}$$

$$0 \le P_i, \qquad \forall i. \tag{1.8}$$

Equations (1.4)–(1.8) are equivalent to

$$\vec{x}^*_{P1} = (P_1, P_2, \ldots, P_{n_n}) = arg\max_{\vec{x}}[\vec{f}^T\vec{x}], \text{ subject to } \mathbf{A}\vec{x} \le \vec{b} \text{ and } \vec{x} \ge 0 \tag{1.9}$$

where

$$A_{kj} = \begin{cases} 1 & \text{if } j = TO(k) \text{ for } k = 1, 2, \ldots, n_e^1, \\ -1 & \text{if } j = FROM(k) \text{ for } k = 1, 2, \ldots, n_e^1, \\ 1 & \text{if } j = TO(k) \text{ for } k = n_e^1 + 1, n_e^1 + 2, \ldots, n_e^1 + n_e^2, \\ \frac{-\alpha_{ji}}{|C(i)|} & \text{if } j = FROM(k) \text{ and } i = TO(k) \text{ for } k = n_e^1 + 1, n_e^1 + 2, \ldots, \\ & n_e^1 + n_e^2, \\ 1 & \text{if node } j \text{ is the } (k - n_e^1 - n_e^2)^{th} \text{ leaf node for} \\ & k = n_e^1 + n_e^2 + 1, \ldots, n_e^1 + n_e^2 + n_L, \\ 0 & \text{else,} \end{cases} \tag{1.10}$$

$$b_k = \begin{cases} \beta_{ji} & \text{if } j = FROM(k) \text{ and } i = TO(k), \\ & \text{for } k = 1, 2, 3, \ldots, n_e^1, \\ 100\left(1 - \sum_{j\in C(i)} \frac{\alpha_{ji}}{|C(i)|}\right) & \text{if } i = TO(k) \text{ for } k = n_e^1 + 1, n_e^1 + 2, \ldots, \\ & n_e^1 + n_e^2, \\ P_i^0 & \text{if node } i \text{ is the } (k - n_e^1 - n_e^2)^{th} \text{ leaf node for} \\ & k = n_e^1 + n_e^2 + 1, \ldots, n_e^1 + n_e^2 + n_L, \end{cases} \tag{1.11}$$

and

$$f_i = \begin{cases} 1 & i = 1, \\ 0 & i = 2, 3, \ldots, n_n, \end{cases} \tag{1.12}$$

where:

n_e^1 is the number of edges in the network associated with the β constraints.
n_e^2 is the number of edges in the network associated with the α constraints.
n_n is the number of nodes.
n_L is the number of leaf nodes.

$C(i)$ is the set of *FROM* nodes going to *TO* node i.

By convention, a *TO* node depends on all of its *FROM* nodes and is the parent of the *FROM* nodes and edge k goes from node *FROM*(k) to node *TO*(k).

1.2.2 A Linear Program to Address Complications Arising from (P1) Having Nonunique Solutions

As will be shown by example, (1.4)–(1.8), or equivalently (1.9), typically have multiple solutions with only one uniquely solving the FDNA equation (1.1). The multiple solutions are the result of the possibility of nonbinding constraints of the optimal solutions of (1.9).

If the objective was $\sum_i P_i$ instead of P_1, the optimization would not have nonbinding constraints. Hence the unique solution to the FDNA equations is $\vec{x}^*_{P2}$, and the FDNA mission objective is $\vec{f}^T \vec{x}^*_{P2}$, where $\vec{x}^*_{P2}$ is the solution to
(P2):

$$\vec{x}^*_{P2} = arg\max_{\vec{x}}[\vec{1}^T \vec{x}], \text{ subject to } \mathbf{A}\vec{x} \leq \vec{b} \text{ and } \vec{x} \geq 0. \tag{1.13}$$

1.2.3 The Dual Linear Program and Its Relationship to the Degradation Index

The dual of (P1) is
(P3):

$$\vec{\lambda}^* = arg\min_{\vec{\lambda}}[\vec{b}^T \vec{\lambda}], \text{ subject to } \mathbf{A}^T \vec{\lambda} \geq \vec{f} \text{ and } \vec{\lambda} \geq 0. \tag{1.14}$$

By construction, the dual finds the marginal impact of the (P1) equations and consequently is a pathway to find the marginal impact and criticality of the operability of individual nodes in the mission network.

First introduced by Servi and Garvey (2017), the Degradation Index uses information about the topology and all of the parameters of the mission network to quantify the marginal impact of a single node's operability on the mission's operability. This measure can be computed using the solution of (P3), $\vec{\lambda}^*$, assuming (P3) has a unique solution. Since $\vec{\lambda}^*$ is the dual solution to the primal problem, (1.4)–(1.8), λ_k^* equals the change of the mission objective from marginally increasing the constant term of the kth constraint of (1.4)–(1.8). However, since the left-hand side of (1.5)–(1.7) is just the operability of a single node, the impact on the mission objective from marginally increasing node i's operability is therefore the maximum of the impact from marginally increasing constant terms of the subset of the (1.5)–(1.7) equations which include the operability of node i on the left-hand side. Hence,

$$\mathcal{D}_i = \max_{\{k: A_{k,i}=1\}} \lambda_k^* = \max_k [\chi(A_{k,i} = 1)\lambda_k^*], \tag{1.15}$$

where $\chi(\cdot)$ is an indicator function equal to 1 if its argument is true and 0 if it is false and λ_k^* is defined in (1.14).

1.2.4 *Addressing Complications When Degradation Index Is not Well Defined*

1.2.4.1 Identifying the Existence of Nonunique Solutions

Unfortunately, a complication might occur: It might be the case that the derivative $\mathcal{D}_i = \frac{dP_1}{dP_i}$ is not defined. As will be shown below, this complication can be detected easily algorithmically. Consider, as a special case of (1.1),

$$P_1 = \min[P_2 + \beta_{ji}, \alpha_{21} P_2 + 100(1 - \alpha_{21})]. \tag{1.16}$$

In this case, $\mathcal{D}_2 = \frac{\partial P_1}{\partial P_2} = 1$ if the first term of (1.16) is the minimum, and $\frac{\partial P_1}{\partial P_2} = \alpha_{21}$ if the second term of (1.16) is the minimum. However, $\frac{\partial P_1}{\partial P_2}$ is undefined if the first and second terms in (1.16) are equal (for the same reason $f(x) = \min(x, 1)$ is not differentiable at $x = 1$).

This complication arises when the dual has a nonunique solution. Detection of nonunique solutions can be accomplished by formulating an optimization problem that seeks to find a solution to the dual problem, (P3), which is farthest from a given known solution, $\vec{\lambda}^*$, using an L1 norm. If (P3) has a unique solution, this will find an L1 norm of 0. Otherwise, it will find a different solution with an L1 norm strictly greater than 0.

$$\Delta = \max_{\vec{\lambda}, \vec{y}, \vec{e}} \quad \vec{1}^T \vec{y} \tag{1.17}$$

subject to

$$\mathbf{A}^T\vec{\lambda} \geq \vec{f}, \tag{1.18}$$

$$b^T\vec{\lambda} \leq b^T\vec{\lambda}^*, \tag{1.19}$$

$$\lambda_i - \lambda_i^* \leq y_i + e_i M, \qquad \forall i, \tag{1.20}$$

$$\lambda_i - \lambda_i^* \geq y_i - (1 - e_i)M, \qquad \forall i, \tag{1.21}$$

$$\lambda_i - \lambda_i^* \leq -y_i + e_i M, \qquad \forall i, \tag{1.22}$$

$$\lambda_i - \lambda_i^* \geq -y_i - (1 - e_i)M, \qquad \forall i, \tag{1.23}$$

$$\vec{\lambda}, \vec{y} \geq 0, \tag{1.24}$$

$$e_i \in \{0, 1\}, \qquad \forall i, \tag{1.25}$$

where M is a large positive number.

One sees the correctness of this formulation as follows: Using the notation of $\vec{\lambda}^\dagger$ being the solution of (1.17)–(1.25), the result of the optimization will find the value of $\vec{\lambda}^\dagger$ which maximizes

$$\Delta = \sum_i |\lambda_i^\dagger - \lambda_i^*|.$$

This can be seen by showing for the following three cases that $y_i = |\lambda_i^\dagger - \lambda_i^*|$:

- *Case 1:* $\lambda_i^\dagger - \lambda_i^* > 0$: If $e_i = 0$, then (1.22) and (1.24) contradict $\lambda_i^\dagger - \lambda_i^* > 0$, as a nonnegative value of y_i forces the smallest possible value of the right-hand side of (1.22) to be 0. Hence, at optimality, $e_i = 1$. This implies that (1.20), (1.22), and (1.23) are nonbinding constraints, which must be true. However, (1.17) and (1.21) imply $y_i = \lambda_i^\dagger - \lambda_i^* = |\lambda_i^\dagger - \lambda_i^*|$.
- *Case 2*: $\lambda_i^\dagger - \lambda_i^* < 0$: If $e_i = 1$, then (1.21) and (1.24) contradict $\lambda_i^\dagger - \lambda_i^* < 0$ because a nonnegative value of y_i forces the largest possible value of the right-hand side of (1.21) to be 0. Hence, at optimality, $e_i = 0$. This implies that (1.20), (1.21) and (1.23) are nonbinding constraints, which must be true. However, (1.17) and (1.22) imply $y_i = \lambda_i^* - \lambda_i^\dagger = |\lambda_i^\dagger - \lambda_i^*|$.
- *Case 3*: $\lambda_i^\dagger - \lambda_i^* = 0$: In this case, (1.20) and (1.23) are true regardless of the value of e_i. If $e_i = 0$, then (1.21) is always true, and (1.22) implies $y_i = 0 = |\lambda_i^\dagger - \lambda_i^*|$. If $e_i = 1$, then (1.22) is always true, and (1.21) implies $y_i = 0 = |\lambda_i^\dagger - \lambda_i^*|$.

Equations (1.17) and (1.25) are equivalent to

(P4):

$$\Delta = \min_{\vec{\lambda}, \vec{y}, \vec{e}} \left[(\vec{0}^T, \vec{1}^T, \vec{0}^T) \begin{pmatrix} \vec{\lambda} \\ \vec{y} \\ \vec{e} \end{pmatrix} \right] \tag{1.26}$$

subject to

$$\begin{pmatrix} \mathbf{A}^T & \mathbf{0} & \mathbf{0} \\ -\vec{b}^T & \vec{0}^T & \vec{0} \\ -\mathbf{I} & \mathbf{I} & M^*\mathbf{I} \\ \mathbf{I} & -\mathbf{I} & -M^*\mathbf{I} \\ -\mathbf{I} & -\mathbf{I} & M^*\mathbf{I} \\ \mathbf{I} & \mathbf{I} & -M^*\mathbf{I} \end{pmatrix} \begin{pmatrix} \vec{\lambda} \\ \vec{y} \\ \vec{e} \end{pmatrix} \geq \begin{pmatrix} \vec{f} \\ -\vec{b}^T\vec{\lambda}^* \\ -\vec{\lambda}^* \\ \vec{\lambda}^* - M^*\vec{1} \\ -\vec{\lambda}^* \\ \vec{\lambda}^* - M^*\vec{1} \end{pmatrix}, \tag{1.27}$$

$$\vec{\lambda}, \vec{y} \geq 0, \tag{1.28}$$

$$e_i \in \{0, 1\}, \qquad \forall i. \tag{1.29}$$

1.2.4.2 Finding Lower Bounds of the Degradation Index if the Dual Has a Nonunique Solution

Servi and Garvey (2017) previously explored the complication of $\mathcal{D}_i$ not being defined. They used an argument that approximated the minimum function in (1.1) as the limit of a sequence of functions which are differentiable everywhere. This leads to the conclusion that a reasonable approximation for $\mathcal{D}_i$ in (1.1) for the case where multiple components of the minimization are equal is to take the average of $\mathcal{D}_i$ over those components. For the example of (1.16), in the case of a tie, Servi and Garvey (2017) would approximate $\mathcal{D}_i$ by $\frac{1+\alpha_{21}}{2}$.

For the case of a nonunique solution to (P3), it is impossible to compute a unique value for the Degradation Index of any of the nodes since, from (1.15), the Degradation Index depends on having a unique value of $\vec{\lambda}^*$. Section 1.2.4 takes a different approach from Servi and Garvey (2017) by instead finding the lower bound of each Degradation Index.

The lower bound of node i's Degradation Index is

$$\mathcal{D}_i^{\text{LOWER}} = \min_{\vec{\lambda}} \mathcal{D}_i^L \tag{1.30}$$

where

$$\mathcal{D}_i^L = \max_{\{k:\lambda_k>0\}} \left[I(A_{ki} = 1)\lambda_k \right] \tag{1.31}$$

subject to

$$\begin{pmatrix} \mathbf{A}^T \\ -\vec{b}^T \end{pmatrix} \vec{\lambda} \geq \begin{pmatrix} \vec{f} \\ -\vec{b}^T\vec{\lambda}^* \end{pmatrix}, \tag{1.32}$$

$$\begin{pmatrix} \vec{\lambda} \\ \mathcal{D}_i^L \end{pmatrix} \geq \vec{0}. \tag{1.33}$$

Equation (1.31) is a modification of (1.15) based on the observation that there is no loss of generality to constrain λ_k to be strictly positive if $\mathcal{D}_i^L$ is constrained to be nonnegative in (1.33). The bottom of (1.32) is an addition to (1.14) which guarantees that $\vec{\lambda}$ is an optimal solution of the dual as $\vec{b}^T\vec{\lambda}^*$ is, by definition, the minimum objective of the dual, $\vec{b}^T\vec{\lambda}$.

Equation (1.30) would have the same solution if the maximization equation in (1.31) were replaced by

$$\mathcal{D}_i^L \geq I(A_{ki} = 1)\lambda_k, \qquad \forall k \text{ with } \lambda_k > 0, \tag{1.34}$$

because the objective is to minimize $\mathcal{D}_i^L$. After substituting (1.34) for (1.31) and defining

$$L_{\ell,k}^i = \begin{cases} 1 & \text{if } k \text{ is the } \ell\text{th value of } \{k : A_{ki} = 1 \text{ and } \lambda_k > 0\}, \\ 0, & \text{else,} \end{cases} \tag{1.35}$$

(1.30) and (1.32)–(1.34) can be put into the matrix form:
(P5):

$$\mathcal{D}_i^{\text{LOWER}} = \min_{\vec{\lambda}, \mathcal{D}_i^L} \mathcal{D}_i^L \tag{1.36}$$

subject to

$$\begin{pmatrix} \mathbf{A}^T & \vec{0} \\ -\vec{b}^T & \vec{0} \\ -\mathbf{L}^i & \vec{1} \end{pmatrix} \begin{pmatrix} \vec{\lambda} \\ \mathcal{D}_i^L \end{pmatrix} \geq \begin{pmatrix} \vec{f} \\ -\vec{b}^T\vec{\lambda}^* \\ \vec{0} \end{pmatrix}, \tag{1.37}$$

$$\begin{pmatrix} \vec{\lambda} \\ \mathcal{D}_i^L \end{pmatrix} \geq \vec{0}. \tag{1.38}$$

1.2.4.3 Finding Upper Bounds of the Degradation Index if the Dual Has a Nonunique Solution

In a similar manner, the upper bound, $\mathcal{D}_i^{\text{UPPER}}$, is defined as the natural variation of (1.30)–(1.33):

$$\mathcal{D}_i^{\text{UPPER}} = \max_{\vec{\lambda}} \mathcal{D}_i^U, \tag{1.39}$$

where

$$\mathcal{D}_i^U = \max_{\{k:\lambda_k>0\}} \Big[I(A_{ki} = 1)\lambda_k \Big], \tag{1.40}$$

subject to

$$\begin{pmatrix} \mathbf{A}^T \\ -\vec{b}^T \end{pmatrix} \vec{\lambda} \geq \begin{pmatrix} \vec{f} \\ -\vec{b}^T\vec{\lambda}^* \end{pmatrix}, \tag{1.41}$$

$$\begin{pmatrix} \vec{\lambda} \\ \mathcal{D}_i^U \end{pmatrix} \geq \vec{0}. \tag{1.42}$$

Unfortunately, it is not possible to frame this as a single linear program without the introduction of binary variables, e_k. To frame such an optimization, one observes that $\mathcal{D}_i^U = \max_{\{k:\lambda_k>0\}}[I(A_{ki} = 1)\lambda_k]$ is equivalent to

$$\mathcal{D}_i^U \geq I(A_{ki} = 1)\lambda_k, \forall k \text{ such that } \lambda_k > 0, \tag{1.43}$$

$$\mathcal{D}_i^U \leq I(A_{ki} = 1)\lambda_k + M(1 - e_k), \forall k \text{ such that } \lambda_k > 0, \tag{1.44}$$

and

$$\sum_k e_k = 1 \tag{1.45}$$

if the objective is to maximize $\mathcal{D}_i^U$, $e_k \in \{0, 1\}$ for all k, and M is a large number. This is seen as (1.43) guarantees that $\mathcal{D}_i^U \geq \max_{\{k:\lambda_k>0\}} \big[I(A_{ki} = 1)\lambda_k\big]$ and (1.44) and (1.45) guarantee that $\mathcal{D}_i \leq \max_{\{k:\lambda_k>0\}} \big[I(A_{ki} = 1)\lambda_k\big]$ if one wants to maximize $\mathcal{D}_i^U$.

The formulation given by (1.39)–(1.42) can also be put in a matrix form using (1.41) and (1.42) and substituting (1.43)–(1.45) for (1.40). Hence,
(P6):

$$\mathcal{D}_i^{\text{UPPER}} = \max_{\vec{\lambda}, \mathcal{D}_i^U, \vec{e}} \mathcal{D}_i^U \tag{1.46}$$

subject to

$$\begin{pmatrix} \mathbf{A}^T & \vec{0} & 0 * \mathbf{I} \\ -\vec{b}^T & 0 & \vec{0}^T \\ -\mathbf{L}^i & \vec{1} & 0 * \mathbf{I} \\ \mathbf{L} & -\vec{1} & -M * \mathbf{I} \\ \vec{0}^T & 0 & \vec{1}^T \end{pmatrix} \begin{pmatrix} \vec{\lambda} \\ \mathcal{D}_i^U \\ \vec{e} \end{pmatrix} \geq \begin{pmatrix} \vec{f} \\ -\vec{b}^T\vec{\lambda}^* \\ \vec{0} \\ -M * \vec{1} \\ 1 \end{pmatrix}, \tag{1.47}$$

$$\begin{pmatrix} \vec{\lambda} \\ \mathcal{D}_i^U \end{pmatrix} \geq \vec{0}, \tag{1.48}$$

$$e_k \in \{0, 1\}, \qquad \forall k, \tag{1.49}$$

where **I** is an identity matrix.

1.3 Examples

Four examples are examined to illustrate key concepts:

- Example 1 has the property that (P1) has a unique solution and therefore solves the FDNA equations. However, its dual has nonunique solutions, so only lower and upper bounds of the Degradation Index can be found.
- Example 2 has the property that (P1) has a nonunique solution, so (P2) is needed to solve the FDNA equations. However, the dual of (P1) has a unique solution, so exact values of the Degradation Index can be found from its dual, (P3).
- Example 3 has the property that (P1) has a unique solution and therefore solves the FDNA equations. Its dual also has a unique solution, so exact values of Degradation Index can be found from (P3). By construction, Example 3 is particularly easy to confirm the correctness of the analysis.
- Example 4 is a large class of missions that permit one to demonstrate that the computation requirements for finding the Degradation Index for realistic missions increase nearly linearly for large networks and that concerns about nonunique dual solutions, which prevent exact calculation of Degradation Index, are more of a theoretical concern than a practical one.

1.3.1 Example 1: A Unique (P1) Solution but a Nonunique Dual

For illustrative purposes, consider the four-node system originally from Figure 1 of Servi and Garvey (2017) which is captured in Table 1.1. This system notational represents the transmission of data from a ground station (node 4) to two relay stations (node 2 and node 3) that uplink the data to a receiving satellite, node 1. The α's and β's in the table reflect the extent to which each node's operability depends on another. For this example, we assume node 4, the Ground Station, has an operability of 25, and the functional dependency parameters of the links are given in Table 1.1.

From (1.1) and Table 1.1 the FDNA equations are given by

$$P_1 = \min[P_3 + 10, P_2 + 10, 0.4P_2 + 0.4P_3 + 20], \tag{1.50}$$

$$P_2 = \min[P_4 + 100, 0.8P_4 + 20], \tag{1.51}$$

$$P_3 = \min[P_4 + 15, 0.75P_4 + 25], \tag{1.52}$$

$$P_4 = 25. \tag{1.53}$$

Equations (1.50)–(1.53) can be interpreted as follows: From (1.51), the parameters of link (4,2) were chosen to illustrate a dependency relationship where the operability of node 2, in the absence of node 4's operability, is 20. From (1.52), the parameters of link (4,3) were chosen to illustrate node 3 having an organic operability of 25, but the operability of node 2 is now capped at only 15, more than the operability of node 4. From (1.50), the parameters of links (2,1) and (3,1) illustrate a low cap on node 1 based on the lesser of the operabilities of node 2 and node 3 plus 10, while having node 1 related to the average operability of node 2 and node 3 if they are high. This example, and two variations of it, will be used to illustrate the above analysis.

Equations (1.50)–(1.53) have the unique solution

$$(P_1^*, P_2^*, P_3^*, P_4^*) = (50, 40, 40, 25). \tag{1.54}$$

This example can be put in the form of (P1), (1.4)–(1.8):

$$\max_{P_1, P_2, P_3, P_4} \quad P_1 \tag{1.55}$$

subject to

$$P_1 \leq P_3 + 10, \tag{1.56}$$

$$P_1 \leq P_2 + 10, \tag{1.57}$$

$$P_2 \leq P_4 + 100, \tag{1.58}$$

$$P_3 \leq P_4 + 15, \tag{1.59}$$

$$P_1 \leq 0.4P_2 + 0.4P_3 + 20, \tag{1.60}$$

$$P_2 \leq 0.8P_4 + 20, \tag{1.61}$$

$$P_3 \leq 0.75P_4 + 25, \tag{1.62}$$

$$P_4 \leq 25, \tag{1.63}$$

or the form of (1.9), where

$$A = \begin{pmatrix} 1 & 0 & -1 & 0 \\ 1 & -1 & 0 & 0 \\ 0 & 1 & 0 & -1 \\ 0 & 0 & 1 & -1 \\ 1 & -0.4 & -0.4 & 0 \\ 0 & 1 & 0 & -0.8 \\ 0 & 0 & 1 & -0.75 \\ 0 & 0 & 0 & 1 \end{pmatrix}, \quad f = \begin{pmatrix} 1 \\ 0 \\ 0 \\ 0 \end{pmatrix}, \quad \text{and} \quad \vec{b} = \begin{pmatrix} 10 \\ 10 \\ 100 \\ 15 \\ 20 \\ 20 \\ 25 \\ 25 \end{pmatrix},$$

which has an optimal solution equal to (1.54). However, at this stage we do not know if this solution is unique. To algorithmically find the unique solution to the FDNA equations, (1.50)–(1.53), following (P2), one needs to instead find

$$\max_{P_1, P_2, P_3, P_4} (P_1 + P_2 + P_3 + P_4) \tag{1.64}$$

subject to (1.56)–(1.63). The solution is (1.54), which is indeed the unique solution to the FDNA equations (1.50)–(1.53).

Identifying criticality measures of nodes requires finding (P3), the dual of (1.55)–(1.63). This corresponds to solving (1.14), or more specifically, solving

$$\min_{s,r,t} \left[10\, s_{31} + 10\, s_{21} + 100\, s_{42} + 15\, s_{43} + 20\, r_1 + 20\, r_2 + 25\, r_3 + 25 t_4 \right] \tag{1.65}$$

subject to

$$s_{31} + s_{21} + r_1 \geq 1, \tag{1.66}$$

$$-\, s_{21} + s_{42} - 0.4\, r_1 + r_2 \geq 0, \tag{1.67}$$

$$-\, s_{31} + s_{43} - 0.4\, r_1 + r_3 \geq 0, \tag{1.68}$$

$$-\, s_{42} - s_{43} - 0.8\, r_2 - 0.75\, r_3 + t_4 \geq 0, \tag{1.69}$$

$$s_{ij}, r_j, t_i \geq 0, \tag{1.70}$$

which has the optimal solution

$$\vec{\lambda}^* = (s_{31}^*, s_{21}^*, s_{42}^*, s_{43}^*, r_1^*, r_2^*, r_3^*, t_4^*) = (1, 0, 0, 1, 0, 0, 0, 1) \tag{1.71}$$

and the optimal objective function $P_1^* = 10 + 15 + 25 = 50$. This suggests, from Eq. (1.15), that

$$\mathcal{D}_2 = \max[s_{42}, r_2] = 0, \quad \mathcal{D}_3 = \max[s_{43}, r_3] = 1, \quad \text{and} \quad \mathcal{D}_4 = \max[t_4] = 1. \tag{1.72}$$

However, this criticality computation is not guaranteed unless the dual has a unique solution. To determine if the dual has a unique solution, one uses (1.17)–(1.25), or equivalently (P4), where $\vec{\lambda}^*$ is given by (1.71). The solution to this optimization is

$$\Delta = 4.2, \quad \vec{\lambda}^\dagger = (0, 1, 0, 0, 0, 1, 0, 0.8), \quad \vec{y}^* = (1, 1, 0, 1, 0, 1, 0, 0.2), \quad \text{and}$$

$$\vec{e}^* = (0, 1, 0, 0, 0, 1, 0, 0).$$

Determining that $\Delta = 4.2 > 0$ signifies that (P3) has a nonunique solution, so (1.72) is not reliable and (P5) and (P6) are needed. Specifically, (P5) is first used to find a lower bound of $\mathcal{D}_i$ for $i = 2$, 3, and 4.

For $i = 2$: From Eqs. (1.36)–(1.38),

$$\mathcal{D}_2^{\text{LOWER}} = \min_{\vec{\lambda}, \mathcal{D}_2} \mathcal{D}_2 \tag{1.73}$$

subject to

$$s_{31} + s_{21} + r_1 \geq 1, \tag{1.74}$$

$$- s_{21} + s_{42} - 0.4\, r_1 + r_2 \geq 0, \tag{1.75}$$

$$- s_{31} + s_{43} - 0.4\, r_1 + r_3 \geq 0, \tag{1.76}$$

$$- s_{42} - s_{43} - 0.8\, r_2 - 0.75\, r_3 + t_4 \geq 0, \tag{1.77}$$

$$\begin{aligned} 10\, s_{31} + 10\, s_{21} + 100\, s_{42} + 15\, s_{43} + 20\, r_1 \\ +20\, r_2 + 25\, r_3 + 25 t_4 \leq 50, \end{aligned} \tag{1.78}$$

$$\mathcal{D}_2 \geq s_{42}, \tag{1.79}$$

$$\mathcal{D}_2 \geq r_2, \tag{1.80}$$

$$s_{ij}, r_j, t_i \geq 0, \qquad \forall i, j, \tag{1.81}$$

where the constraints (1.79) and (1.80) are needed, from (1.36) and (1.38), because $A_{32} = A_{62} = 1$ and s_{42} and r_2 are the third and sixth dual variables. This has the solution $\mathcal{D}_2^{LOWER} = 0$.

To find $\mathcal{D}_3^{\text{LOWER}}$, minimize over $\mathcal{D}_3$ and replace constraints (1.79) and (1.80) with $\mathcal{D}_3 \geq s_{43}$ and $\mathcal{D}_3 \geq r_3$. This has the solution $\mathcal{D}_3^{\text{LOWER}} = 0$.

Finally, to find $\mathcal{D}_4^{\text{LOWER}}$, minimize over $\mathcal{D}_4$ and replace constraints (1.79) and (1.80) with $\mathcal{D}_4 \geq t_4$. This has the solution $\mathcal{D}_4^{\text{LOWER}} = 0.8$.

From (1.46)–(1.49), find the upper bound of $\mathcal{D}_2$, by replacing the objective (1.73) with

$$\mathcal{D}_2^{\text{UPPER}} = \max_{\vec{\lambda}, \mathcal{D}_2} \mathcal{D}_2,$$

subject to the constraints (1.74)–(1.81) along with the constraints

$$\mathcal{D}_2 \geq s_{42} + M(1 - e_1), \tag{1.82}$$

$$\mathcal{D}_2 \geq r_2 + M(1 - e_2), \tag{1.83}$$

$$e_1 + e_2 = 1, \tag{1.84}$$

$$e_i \in \{0, 1\}, \tag{1.85}$$

where M is a large positive number. This has a solution of $\mathcal{D}_2^{UPPER} = 1$. A similar analysis leads to $\mathcal{D}_i^{UPPER} = 1$ also for $i = 3$ and $i = 4$.

Combining the analysis of the lower and upper bounds, one concludes that

$$\mathcal{D}_2 \in [0, 1], \ \mathcal{D}_3 \in [0, 1] \ \text{and} \ \mathcal{D}_4 \in [0.8, 1].$$

1.3.2 Example 2: A Nonunique (P1) Solution but a Unique Dual

If the FDNA equations (1.50)–(1.53) are modified such that (1.53), $P_4 = 25$, is replaced by $P_4 = 20$ then the solution to this modified set of FDNA equations is

$$(P_1^*, P_2^*, P_3^*, P_4^*) = (45, 36, 35, 20). \tag{1.86}$$

If (P1), as given by (1.55)–(1.63), is modified to reflect the above modification of (1.53) by making $P_4 \leq 20$ in (1.63), there are an infinite number of optimal solutions of (P1) including the FDNA solution (1.86). For example, another optimal solution of (P1) is

$$(P_1^*, P_2^*, P_3^*, P_4^*) = (45, 35, 35, 20), \tag{1.87}$$

as this solution also satisfies all the constraints but does not change the objective function, $P_1^* = 45$. However, it does not satisfy the FDNA equations (1.50)–(1.53).

If (P2), as shown in (1.13), is modified to reflect the above modification of (1.53) in order to algorithmically find the unique solution to the FDNA equations, one must modify (P2) by changing (1.63) to $P_4 \leq 20$. This results in the unique solution to the modified (P2) being (1.86), which indeed is the corresponding solution to the modified FDNA equation.

If the dual, (P3), as given in (1.65)–(1.70), is modified to reflect the above modification of (1.53), the objective function, (1.65), replaces the term "$25r_1$" with "$20r_1$," and the optimal solution, after modification, becomes

$$(s_{31}^*, s_{21}^*, s_{42}^*, s_{43}^*, r_1^*, r_2^*, r_3^*, t_4^*) = (1, 0, 0, 1, 0, 0, 0, 1)$$

with the optimal value of the objective function now being $P_1^* = 45$. This suggests, from (1.15), that

$$\mathcal{D}_2 = \max[s_{42}, r_2] = 0.0, \ \mathcal{D}_3 = \max[s_{43}, r_3] = 1.0, \ \text{and } \mathcal{D}_4 = \max[t_4] = 1.0. \tag{1.88}$$

However, this criticality computation is not reliable unless the dual has a unique solution. This can be determined by examining (P4), the optimization problem corresponding to (1.26)–(1.29), or, equivalently, examining (1.17)–(1.25). The solution to this optimization is

$$\Delta = 0.0, \quad \vec{\lambda}^\dagger = (1, 0, 0, 1, 0, 0, 0, 1), \quad \vec{y}^* = (0, 0, 0, 0, 0, 0, 0, 0), \quad \text{and}$$
$$\vec{e}^* = (0, 0, 0, 0, 0, 1, 1, 1).$$

That $\Delta = 0.0$ (or more precisely, Δ is less than the tolerance accuracy of the algorithm) signifies that the solution to (P3) is unique, and therefore (1.88) is correct.

1.3.3 Example 3: A Unique (P1) Solution and a Unique Dual

This example modifies Example 1 by using the α parameters from Table 1.1 but now sets all β terms to 100. Also, this example assumes $P_4 = 25$.

Since all of the values of P_i are less than 100 and all β's equal 100, (1.50)–(1.53) can be simplified to

$$P_1 = 0.4P_2 + 0.4P_3 + 20, \tag{1.89}$$

$$P_2 = 0.8P_4 + 20, \tag{1.90}$$

$$P_3 = 0.75P_4 + 25, \tag{1.91}$$

$$P_4 = 25, \tag{1.92}$$

which implies

$$P_1 = 0.32P_4 + 0.3P_4 + 38 = .62P_4 + 38. \tag{1.93}$$

Putting this example into the framework of (P1), (1.4)–(1.8), after ignoring the β equations as they must all be nonbinding, leads to the optimization problem

$$\max_{P_1, P_2, P_3, P_4} \quad P_1 \tag{1.94}$$

subject to

$$P_1 \leq 0.4P_2 + 0.4P_3 + 20, \tag{1.95}$$

$$P_2 \le 0.8P_4 + 20, \tag{1.96}$$

$$P_3 \le 0.75P_4 + 25, \tag{1.97}$$

$$P_4 \le 25, \tag{1.98}$$

which has the solution

$$P_1^* = 53.5, \qquad P_2^* = 40.0, \qquad P_3^* = 43.75 \qquad P_4^* = 25.0.$$

This is also the solution to the corresponding (P2) problem, so the above solution is also the solution to the FDNA equations. The solution to the dual, (P3), is

$$r_1^* = 1, \qquad r_2^* = 0.4, \qquad r_3^* = 0.4, \qquad t_4^* = 0.62.$$

From (1.15),

$$\mathcal{D}_2 = 0.4, \qquad \mathcal{D}_3 = 0.4, \qquad \mathcal{D}_4 = 0.62. \tag{1.99}$$

However, this computation is not reliable unless the dual has a unique solution. This can be determined by examining (P4), the optimization problem corresponding to (1.27)–(1.29), or, equivalently, examining (1.17)–(1.25). The solution to this optimization is

$$\Delta = 0.0, \quad \vec{\lambda}^{\dagger} = (1, 0, 0, 1, 0, 0, 0, 1), \quad \vec{y}^* = (0, 0, 0, 0, 0, 0, 0, 0), \quad \text{and}$$
$$\vec{e}^* = (0, 0, 1, 1, 1, 0, 0, 0).$$

That Δ is less than the tolerance accuracy of the algorithm signifies that the solution to (P3) is unique, and therefore (1.99) is correct. This should be expected as (1.99) is consistent with (1.89) and (1.93) using the definition, from (1.3) that $\mathcal{D}_i = \frac{dP_1}{dPi}$.

1.3.4 Example 4: Larger Examples

This section will follow the analysis of Servi and Garvey (2017) and examine a sequence of realistic larger examples to both quantify the relationship between the speed of the solution and the size of the problem and document that, although theoretically possible to have multiple solutions to the dual problem (P3), which would interfere with the computation of the critically metrics, it is practically not an issue of concern.

Following the analysis in Servi and Garvey (2017) to quantify the timing of the algorithms in Sect. 1.2, consider a random hierachical network consisting of eight levels, where:

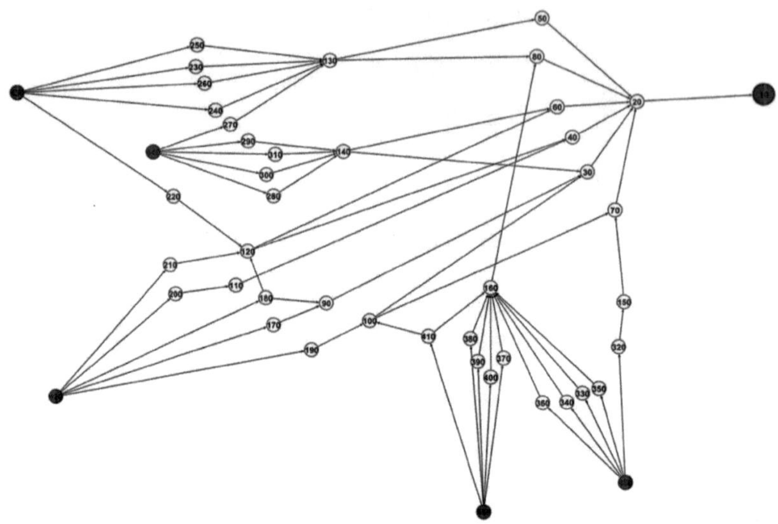

Fig. 1.2 An illustrative example of a randomly generated network

(i) For all but the bottom level, the number of child nodes of each node at the next higher level is first chosen randomly from a geometric distribution with a mean of 5.

(ii) The number of nodes on the lowest level has five nodes.

(iii) Between nodes at adjacent levels, additional links were added, where the number of such links was drawn from a geometric distribution with a mean of 5, and the selected parent and child nodes are randomly selected. These additional links are added, so the network has the property that multiple parent nodes might be dependent on the same child nodes.

Hundreds of such networks were generated. Figure 1.2 is one such network. It was found that the computation time was approximately $-4 \times 10^{-8} N^2 \times 10^{-4} N - 0.03$ seconds for a N node network, i.e., it is nearly linear and less than 0.5 seconds for 2000 node networks.

This near-linearity is related to the observation that the number of paths and the number of links in this class of networks are also a near-linear function of the number of nodes. While this is not true of a purely random network, this property tends to be representative of hierachical networks (or nearly hierarchical networks), which would typically be encountered in realistic settings. The number of paths and the number of links tended to be less than five times the number of nodes. As such, the memory requirements for such networks with less than 10^5 number of nodes are insignificant.

1.4 Conclusion

The importance of building resilience into systems that underpin global missions is more crucial than ever. To enhance resilience against adverse conditions, it is vital to identify critical nodes, whose degradation most impairs the system's operation. However, pinpointing these can be challenging, particularly for large-scale systems. This chapter presents a method to determine and locate globally critical nodes from local dependencies within a system, based on a criticality definition initially proposed by Servi and Garvey (2017). However, it uses a more modern approach based on the dual solution to linear programs. This can help focus system engineers to incorporate more resilience into the system structures. The near-linear computation time and minimal memory requirements of this approach for realistic networks indicate that it is computational feasible for large networks.

References

Bixby RE (2015) Computational progress in linear and mixed integer programming. In: Presentation at ICIAM

Commander, Navy Region Hawaii (CNRH) (2023) Asset management program: Red Hill Administrative Consent Order Section 6 documents (redacted). Commander, Navy Installations Command. https://cnrh.cnic.navy.mil/Portals/79/CNRH/Documents/red_hill/2023%20ACO/Section%206%20Documents/Asset-Management-Program-25SEP2023_Redacted.pdf. See Section 6.5.2.1.1

Franklin B (1795) The way to wealth. Antoine-Augustin Renouard, Paris. https://archive.org/details/bim_eighteenth-century_the-way-to-wealth-or-poo_franklin-benjamin_1795

Garvey PR, Pinto CA (2009) Introduction to functional dependency network analysis. In: Proceedings from the second international symposium on engineering systems. Massachusetts Institute of Technology, Cambridge. http://esd.mit.edu/symp09/day3.html

Garvey PR, Pinto CA (2012) Advanced Risk Analysis in Engineering Enterprise Systems. Chapman-Hall/CRC-Press, Taylor & Francis Group (UK), Boca Raton

Servi LD, Garvey PR (2017) Deriving global criticality conditions from local dependencies using function dependency network analysis (FDNA). Syst Eng 20(4):297–306

Chapter 2
A Continuous Universal Framework for Mission Dependency Networks

Standardization does not mean that we all wear the same color and weave of cloth, eat standard sandwiches, or live in standard rooms with standard furnishings. Homes of infinite variety of are built with a few types of bricks, and with lumber of standard sizes, and with water and heating pipes and fittings of standard dimensions.

W. Edwards Deming (1951)

2.1 Introduction

Chapter 1 introduced one approach, FDNA, to quantitatively capture the interdependencies between a mission's components under the assumption that the mission effectiveness is in [0, 100]. This chapter introduces a simple general framework capable of encapsulating FDNA and nearly all useful dependency representations as special cases. Hence, any future algorithm applied to this standardized near-universal representation of a mission will be applicable to the analysis of missions described using most other representations. This property will be illustrated with a mission network analysis from both an attacker and a defender perspective, using different dependency representations that can be explicitly shown as the special cases of the framework developed in this chapter.

The remainder of the chapter is organized as follows. Section 2.2 formally defines the new framework and illustrates the approach to extending its building blocks to the construction of two commonly used operations: (i) handling a *max* dependency and (ii) handling a *split* dependency. In Sect. 2.3, for illustrative purposes the framework is specialized to capture the dependency relationship of three different functional dependency representations: (i) Functional Dependency Network Analysis (FDNA) (Garvey and Pinto 2009, 2012; Garvey et al. 2014; Servi and Garvey 2017), (ii) System Operational Dependency Analysis (SODA) (Zusack et al. 2018; Guariniello and DeLaurentis 2016; Ubagg 1985), and (iii) Crown Jewel Analysis (CJA) (Hastings et al. 2009; Watters et al. 2009). FDNA is chosen as it is widely used; SODA is chosen to illustrate how a dependency representation,

L. Servi, *Mission Dependency Network Analysis*, Risk, Systems and Decisions,
https://doi.org/10.1007/978-3-032-11260-6_2

which does not naturally fit into the framework, can be approximated with arbitrarily close precision and fit into the framework; CJA is chosen to illustrate that this new framework can also encapsulate a dependency representation whose operability levels and dependency relationships come from finite discrete sets. Section 2.4 states a precise Mixed Integer Linear Program (MILP) formulation of the best actions taken by an attacker and a defender of a mission network using this new framework. In both cases, the analysis is formulated twice: The first formulation assumes that all parameters are known perfectly, and the second formulation handles the more realistic situation where the parameters are known only to be in a known interval; therefore a hedging approach is needed to be robust against known unknowns. Numerical results are presented using the special case of FDNA.

2.2 A New Near-Universal Framework

The new framework uses data elements and variables motivated by FDNA and SODA, which are supplemented by multi-arc dependency relationships between two nodes, negative slope linear functions, separating the Average and Minimum type dependencies, and assuming operability scores in [0, 1] rather than [0, 100].

The new framework is formally defined by the simple Eqs. (2.1) and (2.2): For all $j \in N$, the operability of node j is defined by

$$O_j = \begin{cases} \max\left(\min\left(\frac{1}{|A_j^S|}\sum_{a\in A_j^S} P_a,\ \min_{a\in A_j^C} P_a,\ 1\right),0\right), & A_j \neq \emptyset, \\ L_j & \text{else}, \end{cases} \tag{2.1}$$

where

$$P_a = m_a O_i + b_a, \qquad a = (i, j)_k^t. \tag{2.2}$$

Equations (2.1) and (2.2) are a versatile approach with broader applicability. As will be demonstrated by example, they capture exactly, or with arbitrarily accurate precision, most continuous or discrete alternative dependency representations. Tables 2.1, 2.2, 2.3, and 2.4 define the indices of each dependency arc, the sets of interest, the input parameters, and the key variables, respectively, which will be used to construct FDNA, SODA, and CJA; those definitions specific to each of these dependency relationships are given in the subsequent subsections.

This chapter revolves around (2.1) and (2.2): They will be shown to be quite versatile to capture other frameworks as special cases, and, as such, the analysis of them implicitly is the analysis for most other frameworks.

Equations (2.1) and (2.2) provide a near-universal framework of dependency representations that characterizes mission networks by using individual arcs, which can each be viewed as a very simple, atomic means of representing dependency

Table 2.1 Definitions for indices of arc $a = (i, j)_k^t$ between two tasks

Index	Terminology	Description
i	Child	Index of node whose operability impacts its parent node
j	Parent	Index of node whose operability is impacted by its child node
k	Dependency index	Index for possibly multiple arcs between two tasks
t	Dependency type	Index of the type of relationship between two tasks
$a = (i, j)_k^t$	Dependency	Index of dependency between i and j of type t and index k

Table 2.2 Set definitions

Set	Condition	Description
N	$N = \{0, 1, \ldots, n\}$	Set of nodes corresponding to tasks
T	$T = \{S, C\}$	Types of dependencies
A_j^t	$a = (i, j)_k^t \in A_j^t$	All dependencies of parent j and type t
A_j	$A_j = A_j^S \cup A_j^C$	All dependencies of parent j
A^t	$A^t = \cup_j A_j^t$	All dependencies of type t

Table 2.3 Parameter definitions

Parameter	Condition	Description
L_j	$\forall j \in N : A_j = \emptyset$	Operability of leaf node j
$m_a \in \mathbf{R}$	$\forall a \in \cup_j A_j$	Slope
$b_a \in \mathbf{R}$	$\forall a \in \cup_j A_j$	Y-intercept

Table 2.4 Variable definitions

Variable	Condition	Description
O_i	$i \in N$	Operability of node i
P_a	$a \in \cup_j A_j$	Operability term for dependency a

relationships. Like any base language, a vocabulary needs to be developed to create higher level dependency relationship functions, which can be applied in constructing a dependency network. This section illustrates this higher level construction by providing two simple examples, a MAX function and a SPLIT function, that are used later in this chapter. In the figures provided throughout this chapter, dashed lines represent type S arcs, solid lines represent type C arcs,[1] and the parameters are written as m_a, b_a labels in images beside the links.

MAX

As shown below, the MAX functionality is simple to compose using components of (2.1) and (2.2), so adding a maximum functionality is not needed to be explicitly

[1] The type C and type S dependencies were motivated by the "Strength of Dependency" and "Criticality of Dependency" dependency types originally defined in FDNA (Garvey and Pinto 2009).

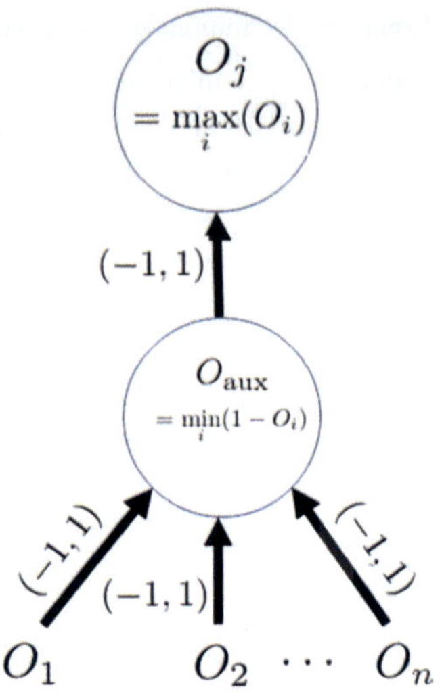

Fig. 2.1 Creating MAX logic for (2.1) and (2.2). Solid line arcs refer to type C dependencies. The pair of parameters on an arc refers to value of m_a followed by b_a. The absence of dashed line arcs indicates that no type S dependencies are needed for this construction

part of the definition of the new framework. Specifically, Fig. 2.1 illustrates how to construct $O_{max} = \max_{i \in I=\{1,2,\ldots,n\}} O_i$ with the introduction of a new node, *aux*, and using only type *C* dependencies denoted by solid line arcs.

Indeed, one can demonstrate the top node's operability $O_{max} = \max_{i \in I} O_i$ as follows. Assume $O_i \in [0, 1]$ for all $i \in I$:

$$\begin{aligned} O_{\mathbf{aux}} &= \max\left(\min\left(\min_{i \in I}[-O_i + 1], 1\right), 0\right) \\ &= \min_{i \in I}[1 - O_i]. \end{aligned}$$

$$\begin{aligned} O_j &= \max\left(\min\left(-O_{\mathbf{aux}} + 1, 1\right), 0\right) \\ &= 1 - O_{\mathbf{aux}} \\ &= 1 - \min_{i \in I}[1 - O_i] \\ &= 1 + \max_{i \in I}[O_i - 1] \\ &= \max_{i \in I} O_i. \end{aligned}$$

This construction of MAX can be used in any place. For convention in drawings, nodes can be drawn as MAX with dotted dependency arcs. The conceptual node MAX will be assumed to define what the real dependencies are.

SPLIT

It is possible to make the following constructed dependency, which is the generalization of a binary decision based on O_i, another node's operability.

$$O_{\mathbf{split}}(O_i, x, y) = \begin{cases} 0, & \text{if } O_i < x, \\ \frac{O_i - x}{y - x}, & \text{if } x \leq O_i < y, \\ 1, & \text{if } y \leq O_i. \end{cases}$$

This function captures the characteristic of dependency relationships where the degradation of operability from 1 has no impact above a threshold, y, before declining, perhaps very quickly, to 0 at x. The SPLIT function provides the basis for a variety of other simple and desirable functions encountered in mission dependency representations. For example, THRESHOLD functions, STEP functions, and IF/THEN/ELSE functions can all be derived from SPLIT very easily. It serves as a building block for creating plateaus on both slides of an operability slope, which represents a rudimentary means to mix discrete logic to a linear analysis.

The SPLIT function is constructed by observing that

$$O_{\mathbf{split}}(O_i, x, y) = \max\left(\min\left(\frac{O_i - x}{y - x}, 1\right), 0\right).$$

Therefore, from (2.1) and (2.2), the construction of SPLIT is done by simply connecting a type C arc from a node corresponding to the operability of i to a node corresponding to $O_{\mathbf{split}}(O_i, x, y)$ with parameters $m_a = 1/(y - x)$ and $b_a = -x/(y - x)$.

2.3 Encapsulation of Other Mission Dependency Representations: Illustrative Special Cases

This section demonstrates how (2.1) and (2.2) encapsulate three different methodologies for mission dependency network analysis: FDNA, SODA, and CJA. FDNA was chosen as it is widely used. Encapsulating SODA illustrates how a framework might need to be approximately with an arbitrarily close precision. CJA illustrates encapsulating a framework whose operability levels and dependency relationships come from a finite discrete set.

2.3.1 Functional Dependency Network Analysis (FDNA) Without OR Nodes

FDNA (Garvey and Pinto 2009, 2012; Garvey et al. 2014) has two types of dependencies: (1) The *Strength of Dependency* (SOD) relationship captures the extent the child's operability impacts the parent's operability, and (2) the *Criticality of Dependency* (COD) captures the most a parent can operate above the operability

Table 2.5 FDNA parameter definitions

Parameter	Condition	Description
$\alpha_a \in [0, 1]$	$a \in A^S$	Strength of Dependency (SOD) parameter
$\beta_a \in \mathbf{R}_+$	$a \in A^C$	Criticality of Dependency (COD) parameter
L_j	$j \notin A_j$	Operability of leaf node j

of a child. The SOD and COD parameters are typically denoted as α_a and β_a, respectively. Using the parameters definitions in Table 2.5, the operability level of a node j is defined by

$$O_j = \begin{cases} \min\left(\frac{1}{|A_j^S|}\sum_{a\in A_j^S} P_a \min_{a\in A_j^C} P_a,\right) & A_j \neq \emptyset, \\ L_j, \quad \text{else}, \end{cases} \quad \forall j \in N \tag{2.3}$$

$$P_a = \alpha_a O_i + (1-\alpha_a)\cdot 100, \qquad \forall a = (i, j) \in A_j^S, \tag{2.4}$$

$$P_a = O_i + \beta_a, \qquad \forall a = (i, j) \in A_j^C, \tag{2.5}$$

$$0 \leq O_j \leq 100, \qquad \forall j \in N, \tag{2.6}$$

where L_j denotes the operability of the leaf node j.

Encapsulating the FDNA Eqs. (2.3)–(2.6) into (2.1) and (2.2) entails (i) a trivial conversion of associating type S arcs with SOD dependencies, (ii) associating type C arcs with COD dependencies, and (iii) scaling down operabilities by a factor of 100. In particular, the operability of node j, using FDNA if j is not an OR node, is given by (2.1) and (2.2), where

$$P_a = \begin{cases} \alpha_a O_i + (1-\alpha_a), & \forall a = (i, j) \in A_j^S, \\ O_i + \frac{\beta_a}{100}, & \forall a = (i, j) \in A_j^C. \end{cases} \tag{2.7}$$

2.3.2 *FDNA Analysis with OR Nodes*

A variation of FDNA, *FDNA with OR nodes*, was introduced (Servi and Garvey 2017) to capture a related dependency relationship. The nodes keep the basic FDNA dependency structure on arcs, but with the introduction of a MAX function. This maximization function has a role not different from an OR node in a reliability network. The operability of node j, if the nodes are all FDNA with OR nodes, is given by

$$O_j = \max_{j\in A_j}\left[\min\left(\alpha_{ij} O_i + (1-\alpha_{ij}), O_i + \beta_{ij}\right)\right].$$

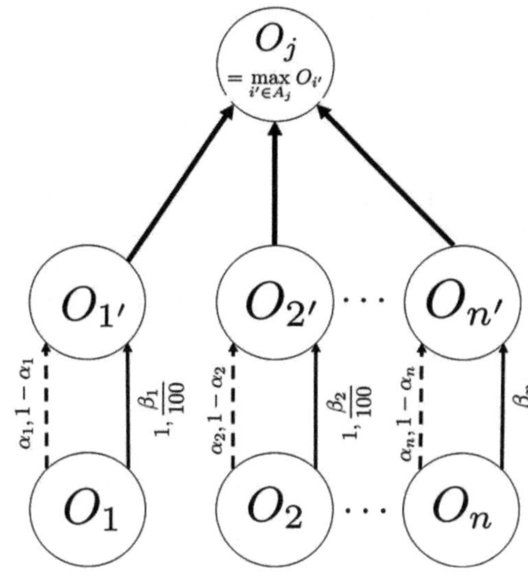

Fig. 2.2 Creating OR logic to extend FDNA

This can be encapsulated using (2.1) and (2.2) by introducing a set of nodes, 1',2', ... associated with nodes 1,2, ..., respectively:

$$O_{i'} = \min[\alpha_{ij} O_i + (1 - \alpha_{ij}), O_i + \beta_{ij}], \tag{2.8}$$

$$O_j = \max_{i' \in A_j} (O_{i'}), \tag{2.9}$$

$$P^S_{i,i'} = \alpha_{ij} O_i + 1 - \alpha_{ij}, \qquad (i, j) \in A_j, \tag{2.10}$$

$$P^C_{i,i'} = O_i + \frac{\beta_{ij}}{100}, \qquad (i, j) \in A_j, \tag{2.11}$$

as illustrated in Fig. 2.2, which uses the MAX construction previously derived.

2.3.3 *System Operational Dependency Analysis (SODA): An Example of Approximating a Dependency Framework*

SODA (Zusack et al. 2018; Guariniello and DeLaurentis 2016) is an example of a dependency representation with a complex continuous function that (2.1) and (2.2) can encapsulate. The purpose of the construction in this section is to demonstrate that (2.1) and (2.2) are versatile enough to capture complex dependency representations. Practically, it would be advantageous to modify SODA to make it less complex while maintaining the subtle advantages that it has.

Table 2.6 SODA parameter definition

Parameter	Condition	Description
L_j	$\forall j \in N$	Operability of leaf node j
G_j	$\forall j \in N$	Ground Operability of node j
$\alpha_a \in [0, 1]$	$a \in A^S$	Strength of Dependency (SOD) term from FDNA
$\beta_a \in [0, 100]$	$a \in A^C$	Criticality of Dependency (COD) term from FDNA
$\gamma_a \in [1, 100]$	$a \in A^C$	Impact of Dependency (IOD) term
$\lambda \in [0, 1]$		Group criticality term, typically positive and close to 0
w_k	$k \in \{1, 2, \ldots, h\}$	Value of W_a where the approximation of $(W_a)^\lambda$ is exact

Table 2.7 SODA variable definition

Variable	Condition	Description
W_a	$a = (i, j) \in A_j^C$	Average operability of siblings
$O_{i,j}^{\mathbf{W}^\lambda}$	$(i, j) \in A_j^C$	Linear approximation of $(W_{i,j})^\lambda$ when $(W_{i,j})^\lambda \in [w_k, w_{k-1})$

SODA may be viewed as an augmented form of FDNA, where (2.12)–(2.13) replace (2.4)–(2.6). It uses FDNA parameters but adds some additional parameters as noted in Tables 2.6 and 2.7.

In the second term of (2.12), SODA introduces a Group Criticality term, $W_{i,j}$, raised to the λ power where λ, the Impact of Dependency (IOD) term, is typically positive and very close to 0. The motivation for including this second term is to guarantee that a parent will have an operability of 0 if all its children have an operability of 0.

$$P_a = \begin{cases} \alpha_a O_i + (1 - \alpha_a) G_j, & \forall a \in A_j^S, \\ \frac{100}{\gamma_a} O_i + (W_{i,j})^\lambda (100 - \beta_a), & \forall a = (i, j) \in A_j^C. \end{cases} \tag{2.12}$$

$$W_{i,j} = \begin{cases} \frac{1}{100} \sum_{(i',j) \in A_j^C \setminus (i,j)} \frac{1}{|A_j^C| - 1} O_{i'}, & |A_j^C| > 1, \\ 1 & |A_j^C| = 1, \end{cases} \quad \forall (i, j) \in A_j^C. \tag{2.13}$$

Other than the Group Criticality term, the conversion to SODA is straightforward, with the only consideration being to normalize the equations so that the operability is in [0,1] rather than [0,100]. For type S dependencies, the construction is

$$P_a = \alpha_a O_i + (1 - \alpha_a) \frac{G_j}{100}, \ \forall a = (i, j) \in A_j^S.$$

As indicated in (2.13), two separate cases must be considered: **Case 1**: $|A_j^C| = 1$ and **Case 2**: $|A_j^C| > 1$.

Case 1: $|A_j^C| = 1$. This is a straightforward case. For type C dependencies, if node j is such that $|A_j^C| = 1$, then

$$P_a = \tfrac{100}{\gamma_a} O_i + 1 - \tfrac{\beta_a}{100} \; \forall a = (i, j) \in A_j^C : |A_j^C| = 1.$$

Case 2: $|A_j^C| > 1$. If node j is such that $|A_j^C| > 1$, the type C dependency for each node i is more involved.

For this case, the construction requires approximating $(W_{i,j})^\lambda$ with a piecewise-linear function, which is exact for $W_{i,j} = w_k, k = 0, 1, \ldots, h$, where $w_0 < w_1 < \ldots < w_h$, $w_0 = 0$ and $w_h = 1$, and will be represented by an auxiliary node for each arc whose operability, $O_{i,j}^{\mathbf{W}^\lambda}$, is set to this approximation. This is done in (2.14) and will require the SPLIT function introduced in Sect. 2.2. Next, (2.14) is used in (2.16), to linearly approximate (2.12). These equations, combined with (2.12), (2.13), and (2.1), define the needed construction that must be used for each j such that $|A_j^C| > 1$.

$$O_{i,j}^{\mathbf{W}^\lambda} = \left[\frac{W_{i,j} - w_{k-1}}{w_k - w_{k-1}} (w_k)^\lambda + \left(1 - \frac{W_{i,j} - w_{k-1}}{w_k - w_{k-1}}\right) (w_{k-1})^\lambda \right]$$
$$if \; W_{i,j} \in [w_k, w_{k-1}), \tag{2.14}$$

$$P_a^{\mathbf{app}} = \max\left(\min\left(\left(\frac{100}{\gamma} O_i + \frac{1-\beta}{100} (W_{i,j})^\lambda\right), 1\right), 0\right), \tag{2.15}$$

$$\approx \frac{100}{\gamma} O_i + \frac{1-\beta}{100} O_{i,j}^{\mathbf{W}^\lambda}, \quad \forall a = (i, j) \in A_j^C. \tag{2.16}$$

The piecewise-linear approximation in (2.14) is relatively simple, but since it must be repeated for every dependency where $|A_j^C| > 1$, it incurs a lot of overhead.

While this construction demonstrates the versatility of (2.1) and (2.2) in approximating nearly any dependency representation, it would be more practical to not use SODA as defined, but instead replace the $O_{i,j}^{\mathbf{W}^\lambda}$ node with a $SPLIT(O_{i,j}, 0, \epsilon)$: $\epsilon > 0$, which achieves a similar goal of guaranteeing that the parent node has an operability of 0 if all its dependency child nodes have an operability of 0. Using a SPLIT node would entail significantly less computational overhead.

2.3.3.1 Crown Jewel Analysis (CJA): An Example of Mapping a Discrete Dependency Framework

In addition to encapsulating *continuous* dependency representations such as FDNA, (2.1) and (2.2) can also be applied to dependency relationships and operability levels that are discrete. This is illustrated by CJA (Watters et al. 2009) using an *Atom*

Mapping approach, which represents each of the discrete operability levels with two binary nodes.

In CJA, there are only four discrete operability levels:

$$O_j \in \{\text{Failure}(fl), \text{Degraded}(dg), \text{Work} - \text{around}(wk), \text{NoImpact}(ni)\},$$

which have a well-defined ordering of

$$fl < dg < wk < ni. \tag{2.17}$$

For each arc, there are three dependency types:

$$t_a \in \{\text{Failure}(FL), \text{Degraded}(DG), \text{Work} - \text{around}(WK)\}.$$

These can be thought of as three different types of the rollup rules in CJA. Here, the dependency types are used to determine the operability level of a parent node according to (2.18) and (2.19). Equation (2.18) defines how child operability level and dependencies are used to compute the parent operability level, and (2.19) shows the impact of multiple child nodes on a single parent.

$$P_a = \begin{cases} fl, & O_i \in \{fl, dg\}, t_a = FL, \\ dg, & O_i \in \{fl, dg\}, t_a = DG, \\ wk, & O_i \in \{fl, dg\}, t_a = WK, \\ ni, & \text{else}, \end{cases} \qquad \forall a \in A, \tag{2.18}$$

$$O_j = \begin{cases} \min_{a \in A_j} P_a, & A_j \neq \emptyset, \\ L_j & else, \end{cases} \qquad \forall j \in N, \tag{2.19}$$

where L_j is the operability level of leaf node j, and the minimum in (2.19) is according to (2.17).

For CJA, the operability level of each node will be represented by two binary nodes as indicated in Table 2.8. For example, node j with operability level *Degraded (dg)* is represented by $O_j^1 = 0$ and $O_j^2 = 1$, according to the second row of Table 2.8.

The binary node mapping approach is described for a simplified version of CJA before being examined in full generality. This will exemplify two different ways of developing the approach for any discrete dependency framework. For the simplified version, the approach entails first representing the operability of each node as one of four levels, which can be encoded with two binary nodes. Next, a table is created listing all possibilities that relate these nodes to the encoding of the resulting parent node operability. For the full generality of CJA, the dependency relationships are presented for the binary nodes; a proof of validation is then provided. For both the simplified and general cases, the relationships can be easily translated into (2.1) and (2.2).

Table 2.8 Encoding CJA operability levels with binary nodes

O_k^1	O_k^2	Operability level
0	0	Failure (fl)
0	1	Degraded (dg)
1	0	Work-around (wk)
1	1	No impact (ni)

Table 2.9 Valuation of all possibilities for parent node j with all three dependency link types

O_{FL}^1	O_{DG}^1	O_{WK}^1	O_j^1	O_j^2	Operability level of parent node j
1	1	1	1	1	No impact (ni)
1	1	0	1	0	Work-around (wk)
1	0	1	0	1	Degraded (dg)
1	0	0	0	1	Degraded (dg)
0	1	1	0	0	Failure (fl)
0	1	0	0	0	Failure (fl)
0	0	1	0	0	Failure (fl)
0	0	0	0	0	Failure (fl)

The Simplified Case For the simplified case, where all parent nodes have at most one dependency type of each type, Table 2.9 encodes the CJA dependency of (2.18) and (2.19) using the coding in Table 2.8. Here, we assume nodes *FL*, *DG*, and *WK* correspond to the dependent children of node j according to $t_a = FL$, $t_a = DG$, and $t_a = WK$, respectively. In Table 2.9, O_{FL}^1, O_{DG}^1, and O_{WK}^1 are the left atoms from Table 2.8 for the FL, DG, and WK node's operability, respectively, and O_j^1 and O_j^2 are the two atoms from Table 2.8 of the parent node j and are derived from (2.18) and (2.19). Note from Table 2.8 and (2.18) that O_{FL}^2, O_{DG}^2, and O_{WK}^2 are not needed to evaluate (2.18). Hence they are not listed in Table 2.9.

Table 2.9 can be illustrated with an example. Suppose parent node j has three dependent children: The first dependent child has operability level wk and has a dependency link type FL, the second dependent child has operability level ni and has a dependency link type DG, and the third dependent child has operability level fl and has a dependency link type WK. Using Table 2.8, since the dependent child related to parent node j was of type FL and has operability level wk, the state of this dependent children is encoded as $O_{FL}^1 = 1$. Similarly, for this example $O_{DG}^1 = 1$, and $O_{WK}^1 = 0$. From (2.18), the value of P_a corresponding to the three dependent children is ni, ni, and wk, respectively. From (2.19), the operability level of the parent node j is wk since, from (2.17), $wk < ni$. In Table 2.8 this is encoded as $O_j^1 = 1$ and $O_j^2 = 0$. This example is captured in the second row of Table 2.9.

For the simplified case, where all parent nodes have at most one dependency type, it is a simple matter to encode the dependency relationship into the form of (2.1) and (2.2). Specifically, from Table 2.9, one can verify that for all cases,

$$O_j^1 = \min\left(O_{FL}^1, O_{DG}^1\right), \tag{2.20}$$

and

$$O_j^2 = \begin{cases} 0, & O_{FL}^1 = 1, O_{WK}^1 = 0, O_j^1 = 1, \\ O_{FL}^1, & \text{else}, \end{cases} \tag{2.21}$$

as $O_j^2 = O_{FL}^1$ except in the second row of Table 2.9.

The General Case For the more general case of having multiple dependency links of the same type from the same parent, the governing (2.20) and (2.21) must be generalized.

To do so, we introduce the notation that $i \in A_j^{FL}$, if node i is a child of node j, and the rollup rule between i and j is FL, i.e., $t_{(i,j)} = FL$. Similarly, we define $i \in A_j^{WG}$ if $t_{(i,j)} = WG$, and $i \in A_j^{DG}$ if $t_{(i,j)} = DG$.

The general case necessitates the introduction of the new variable

$$O_i^{WK} = \begin{cases} 0 & O_j^1 = 1 \text{ and } O_i^1 = 0, \\ 1 & \text{else}. \end{cases} \qquad \forall i \in A_j^{WK}. \tag{2.22}$$

This is equivalent to

$$O_i^{WK} = \max\left(\min\left(\frac{1}{2}\left[-2O_j^1 + 2 + 2O_i^1\right], 1\right), 0\right), \qquad \forall i \in A_j^{WK}. \tag{2.23}$$

As will be argued below, the generalization of (2.20) is

$$O_j^1 = \begin{cases} \min_{i \in (A_j^{FL} \cup A_j^{DG})} O_i^1, & A_j^{FL} \cup A_j^{DG} \neq \phi, \\ 1, & \text{else}, \end{cases} \tag{2.24}$$

and the generalization of (2.21) is

$$O_j^2 = \min\left[\min_{i \in A_j^{FL}} O_i^1, \min_{i \in A_j^{WK}} O_i^{WK}\right]. \tag{2.25}$$

Equations (2.23), (2.24), and (2.25) provide the basis using (2.1) and (2.2) to represent CJA. The validity of (2.24) and (2.25) can be demonstrated by examining the following three mutually exclusive cases that capture all possible dependencies of node j:

1. There is an $i \in A_j^{FL}$ with $O_i^1 = 0$.
2. If $i \in A_j^{FL}$, $O_i^1 = 1$; if $A_j^{DG} \neq \phi$, there is an $i' \in A_j^{DG}$ with $O_{i'}^1 = 0$.
3. If $i \in A_j^{FL}$, $O_i^1 = 1$; if $A_j^{DG} \neq \phi$, $O_{i'}^1 = 1$ for all $i' \in A_j^{DG}$.

Case 1: There is an $i \in A_j^{FL}$ with $O_i^1 = 0$.

From Table 2.8 and the definition of A_j^{FL}, there is at least one child such that $t_{(i,j)} = FL$ and $O_i \in \{fl, dg\}$. From (2.17) and the first equation in (2.18), there is therefore at least one dependency, a, to parent node j such that $P_a = fl$. Hence from (2.19), $O_j = fl$, which, from Table 2.8, encodes to $O_j^1 = 0$ and $O_j^2 = 0$. Since $i \in A_j^{FL}$ and $O_i^1 = 0$, this is consistent with (2.24), which also implies $O_j^1 = 0$. Also, $O_j^2 = 0$ is consistent with (2.25), since $i \in A_j^{FL}$ and $O_i^1 = 0$ implies $\min_{i \in A_j^{FL} \cup A_j^{DG}} O_i^1 = 0$, and, from (2.25), $O_j^2 = 0$ irrespective of the values of any O_i^{WK} nodes.

Case 2: If $i \in A_j^{FL}$, $O_i^1 = 1$; if $A_j^{DG} \neq \phi$, there is an $i' \in A_j^{DG}$ with $O_{i'}^1 = 0$.

From the assumptions of this case, Table 2.8, and the definition of A_j^{FL} and A_j^{DG}, if there is a child $i \in A_j^{FL}$, $O_i \in \{wk, ni\}$. So, from (2.17)–(2.19), $O_j \neq fl$.

If $A_j^{DG} = \phi$, from (2.17)–(2.19), this leads to $P_a \geq wk$ for all a, and $O_i \in \{wk, ni\}$, i.e., $O_j^1 = 1$. Hence, there are any $i \in A_j^{WK}$, then $O_i^{WK} = 1$ from (2.23) since by assumption $O_i^i = 1$. This is consistent with (2.24). However, if $O_i \in \{wk, ni\}$ for all $i \in A_j^{WK}$, then from (2.18), $P_a = ni$ for all a so $O_j = ni$ and $O_j^2 = 1$. This is consistent with (2.25) as $O_i^1 = 1$ for all $i \in A_j^{FL}$ and $A_j^{DG} = \phi$.

On the other hand, if $A_j^{DG} \neq \phi$, there is at least one child, i', such that $i' \in A_j^{DG} \neq \phi$ and $O_{i'} \in \{fl, dg\}$. From (2.18), there is therefore at least one dependency, a, to parent node j such that $P_a = dg$ and no dependency a' to parent node j such that $P_{a'} = fl$. The latter means $O_j \neq fl$, so $O_j \geq dg$. Hence, from (2.19), $O_j = dg$, which, from Table 2.8, encodes to $O_j^1 = 0$ and $O_j^2 = 1$. Since there is at least one $i' \in A_j^{DG} \neq \phi$ with $O_{i'}^1 = 0$ from this case's assumption, this is consistent with (2.24), which also implies $O_j^1 = 0$. Since $O_j^1 = 0$, if there are one or more type WK dependencies to node j, from (2.23), the corresponding value of O_i^{WG} would be 1. Hence, both terms of (2.25) are 1, and $O_j^2 = 1$ is consistent with (2.25).

Case 3: If $i \in A_j^{FL}$, $O_i^1 = 1$; if $A_j^{DG} \neq \phi$, $O_{i'}^1 = 1$ for all $i' \in A_j^{DG}$.

In this case, from Table 2.8 and the definition of A_j^{FL} and A_j^{DG}, if there is a child $i \in A_j^{FL} \cup A_j^{DG}$, then $O_i \in \{wk, ni\}$. Therefore, from (2.18), for all dependencies, a, to parent node j, $P_a \in \{wk, ni\}$. Hence, from (2.19), $O_j \in \{wk, ni\}$. From Table 2.8, this encodes to $O_j^1 = 1$. Hence, if $i \in A_j^{FL} \cup A_j^{DG}$, $O_i^1 = 1$. This is consistent with (2.24), which also implies $O_j^1 = 1$. Since $O_j^1 = 1$, if there are one or more type WK dependencies to node j, from (2.23), the corresponding value of O_i^{WG} would be 1. Hence, both terms of (2.25) are 1, and $O_j^2 = 1$ is consistent with (2.25).

2.4 Analysis

This section describes how to determine the best coordinated attack or defense of a mission network that is characterized by the framework defined by (2.1) and (2.2). For the attack formulation, it is assumed that only the leaf nodes could be attacked, and such attacks would lower their operability with propagating impacts that degrade the mission operability. The attacker seeks to plan an attack that minimizes the mission operability subject to an assumed budget constraint, which limits the number of such attacks. In contrast, for the defense formulation, it is assumed that any leaf node not defended will be attacked. Here the defender seeks to defend against attacks on leaf nodes to maximize the mission operability, subject to an assumed budget constraint that limits the number of such defensive mitigations. Both the attack formulation and the defender formulation are of the form of a Mixed Integer Linear Program (MILP) which can be readily solved using standard commercial software.

First, both analyses are performed assuming that all parameters are known precisely. This is followed by a robust optimization analysis, where it is assumed that the parameters are only known to be within a specific range. Here, the best attack or defense hedges against this imperfect knowledge. For such an analysis to be well defined, it is necessary to make mathematically precise assumptions about the desired extent that one guards against this imperfect knowledge. This might be viewed as the user's level of conservatism. In Sect. 2.4.4 the precise mathematical definition of conservatism will be introduced. Since this analysis is developed for (2.1) and (2.2), any functional dependency representation that can be fit into this framework, either exactly or approximately, can be examined easily as a special case.

2.4.1 Attacker Formulation: No Robustness

The optimization program in (2.26)–(2.38) formulates an attacker's decision process and its impact on the mission network operability. This formulation assumes that the leaf nodes of the mission network are the only nodes that can be accessed for manipulation or attack.[2] When those nodes are attacked, it is assumed that their operability levels drop from L_i to ℓ_i. This operability loss is assumed to propagate upward through the network to ultimately impact the mission operability represented by node 0 and denoted by O_0. This loss propagation follows (2.1) and (2.2). The attacker is assumed to have a resource budget, B, which influences its most impactful viable coordinated attack to reduce the mission operability. The resource budget might be as simple as setting the number of nodes that could

[2] This assumption could be loosened with minor changes to the optimization formulation.

Table 2.10 Definitions for non-robust optimization formulation

Parameter	Condition	Description
O_j	$j \in N$	Operability of node j; if $j = 0$: operability of mission
R_j^t	$j \in N : A_j \neq \emptyset; t \in \{S, C\}$	Operability contribution for parent j from type t dependency
I_j^{O1},I_j^{O2},I_j^{O3}	$\forall j \in N : A_j \neq \emptyset$	Binary variable indicating which part of (2.1) dominates
M		A large positive constant.
$I_{(i,j)}^C = 1$	$i \in A_j^C$	Indicator of dominant type C dependency if all $(i, j) : i \in A_j^C$
P_a	$a = (i, j)_k^t$;	Contribution of dependency (i, j) of type t and index k
m_a, b_a	$a = (i, j)_k^t$;	Parameters to define P_a
z_j	$\forall j \in N : A_j = \emptyset$	The binary variable indicating whether leaf node i is attacked
L_j	$\forall j \in N : A_j = \emptyset$	Operability of leaf node j if not attacked
ℓ_j	$\forall j \in N : A_j = \emptyset$	Operability of leaf node j if attacked
c_j	$\forall j \in N, A_j = \emptyset$	The relative cost to attack node j
B		Attacker's resource budget

be attacked, or it may include the relative cost (in time, funds, or lost future opportunities) of attacking different nodes. The optimization is not robust in the sense that it relies heavily on the assumption that all parameters are known precisely, i.e., it is not hedging against any unknowns. Table 2.10 lists the notation needed for the optimization program.

Equations (2.27)–(2.34) ensure that (2.1) and (2.2) are followed: Specifically, (2.27)–(2.30) capture the minimization function in (2.1). Equations (2.31)–(2.33) encode the components of the two different dependency types in (2.1). Equation (2.34) encodes (2.2). Equation (2.36) adds the attacker's resource budget constraint and (2.35) the impact of the attacker compromising leaf node j.

2.4.2 *Attacker Formulation with Robustness*

With the formulation of the previous section, the user can set the resource budget B, and the optimization will find the combination of nodes to attack that will minimize the mission operability under the resource constraint. Since this was applied to the framework of (2.1) and (2.2), it can address this problem for any dependency representation assumptions which can be represented by (2.1) and (2.2). However, without considering the uncertainty inherent in the dependency parameters in (2.1) and (2.2), the calculated solution does not hedge against known unknowns. This lack of robustness is addressed below in (2.40)–(2.61).

$$\min_{\vec{z}} [O_0] \tag{2.26}$$

subject to

$$O_j \geq R_j^S - (1 - I_j^{O1})M, \qquad \forall j \in N : A_j \neq \emptyset, \tag{2.27}$$

$$O_j \geq R_j^C - (1 - I_j^{O2})M, \qquad \forall j \in N : A_j \neq \emptyset, \tag{2.28}$$

$$O_j \geq 1 - (1 - I_j^{O3})M, \qquad \forall j \in N : A_j \neq \emptyset, \tag{2.29}$$

$$1 = I_j^{O1} + I_j^{O2} + I_j^{O3}, \qquad \forall j \in N : A_j \neq \emptyset, \left|A_j^C\right| > 0, \tag{2.30}$$

$$R_j^S \geq \frac{1}{|A_j^S|} \sum_{i \in A_j^S} P_{(i,j)}^S, \qquad \forall j \in N : A_j \neq \emptyset, \left|A_j^S\right| > 0, \tag{2.31}$$

$$R_j^C \geq P_{(i,j)}^C - (1 - I_{(i,j)}^C)M, \qquad \forall j \in N, i \in A_j^C, \tag{2.32}$$

$$\sum_{i \in A_j^C} I_{(i,j)}^C = 1, \qquad \forall j \in N : \left|A_j^C\right| > 0, \tag{2.33}$$

$$P_a \geq m_a O_i + b_a, \qquad \forall a \in A, \tag{2.34}$$

$$O_j \geq L_j - z_j(L_j - \ell_j), \qquad \forall j \in N : A_j = \emptyset, \tag{2.35}$$

$$\sum_{j \in N} c_j z_j \leq B, \tag{2.36}$$

$$O, P, R \geq 0, \tag{2.37}$$

$$I, z \in \{0, 1\}. \tag{2.38}$$

Specifically, the optimization program (2.40)–(2.59) finds the best coordinated attacker decisions to mostly impact the mission network operability, O_0, while being robust to the uncertainty of parameters, which characterize the dependency relationships. This formulation assumes that the leaf nodes of the mission network are the only nodes that can be accessed for manipulation or attack. To model robustness, we introduce a realization parameter, Γ_{tot}, which dictates the total number of parameters not equal to their nominal value, but instead equal to the value within a specific range which is most disadvantageous to the attacker. Setting Γ_{tot} to 0 corresponds to the non-robust analysis of the previous section. Setting Γ_{tot} to the number of parameters corresponds to the assumption that all parameters have the value within a specified interval which would mostly undermine the attacker's goals. Typically this is an excessively conservative assumption and is not recommended.

Instead, this model allows the user to use Γ_{tot} to select a level of conservatism deemed appropriate and then evaluate decisions that are robust against the worst-case scenario given that chosen level of conservatism. In technical terms, the model is a maximin optimization program which minimizes over attack decisions and

Table 2.11 Definitions for robust optimization formulation

Parameter	Condition	Description
Δ_a^m, Δ_a^b	$a = (i, j)_k^t$;	Possible Change in m_a and b_a used to define P_a
Δ_j^L	$\forall j \in N : A_j = \emptyset$	Possible Change of leaf operability of node j if not attacked
I_a^m, I_a^b	$\forall a \in A$	Indicates uncertainty realized for m_a and b_a
I_j^L	$j \in N : A_j = \emptyset$	Indicates uncertainty realized if node j is not attacked
Γ^G,Γ^{mS},Γ^{mC},,Γ^{bS},Γ^{bC}		Max. number of uncertainty realized of each category
Γ_{tot}		Maximum number of uncertainty realized of all categories
U_a^m	$\forall a \in A$	The proxy for the product of O_i and I_a^m

maximizes over which Γ_{tot} parameters are selected that do not equal their nominal values. This maximization also determines the value of these selected parameters.

The formulation uses the definitions from Table 2.10 supplemented with the definitions from Table 2.11.

Equations (2.40)–(2.59) can be understood as follows: In the expanded formulation, (2.41)–(2.47) are identical to (2.27)–(2.33), and (2.59) is identical to (2.36). Constraints (2.50)–(2.52) force

$$U_a^m = \begin{cases} O_i, & I_a^m = 1, \\ 0, & I_a^m = 0, \end{cases}$$

which is equivalent to the nonlinear equation

$$U_a^m = O_i I_a^m. \tag{2.39}$$

From (2.39), (2.48) is a generalization of (2.34), where the slope of P_a is $m_a + I_a^m \widetilde{m}_a O_i$, and the intercept is $b_a + I_a^b \widetilde{b}_a$. Similarly, (2.49) is a generalization of (2.35), where I_i^L and z_i control leaf node i's operability. Finally, (2.53)–(2.58) encode the number and type of parameters which can be set to values most disadvantageous to the attacker. These equations could be combined into a single equation, but the separation leaves open the option of constraining some types of parameters more than others.

$$\max_{\vec{I}^m, \vec{I}^b, \vec{I}^G} \left[\min_{\vec{z}} \left[O_0 \right] \right] \tag{2.40}$$

subject to

$$O_j \geq R_j^S - (1 - I_j^{O1})M, \qquad \forall j \in N : A_j \neq \emptyset, \tag{2.41}$$

$$O_j \geq R_j^C - (1 - I_j^{O2})M, \qquad \forall j \in N : A_j \neq \emptyset, \tag{2.42}$$

$$O_j \geq 1 - (1 - I_j^{O3})M, \qquad \forall j \in N : A_j \neq \emptyset, \tag{2.43}$$

$$1 = I_j^{O1} + I_j^{O2} + I_j^{O3}, \qquad \forall j \in N : \left|A_j^C\right| > 0, \tag{2.44}$$

$$R_j^S \geq \frac{1}{|A_j^S|} \sum_{i \in A_j^S} P_{(i,j)}^S, \qquad \forall j \in N : \left|A_j^S\right| > 0, \tag{2.45}$$

$$R_j^C \geq P_{(i,j)}^C - (1 - I_{(i,j)}^C)M, \qquad \forall j \in N, i \in A_j^C, \tag{2.46}$$

$$\sum_{i \in A_j^C} I_{(i,j)}^C = 1, \qquad \forall j \in N : \left|A_j^C\right| > 0, \tag{2.47}$$

$$P_a \geq m_a O_i + \Delta_a^m U_a^m + b_a + I_a^b \Delta_a^b, \qquad \forall a \in A, \tag{2.48}$$

$$O_i \geq L_i + I_i^L \Delta_i^L - z_i(L_i - \ell_i), \qquad \forall i \in N : A_i = \emptyset, \tag{2.49}$$

$$U_a^m \leq I_a^m, \qquad \forall a \in A, \tag{2.50}$$

$$U_a^m \leq O_i, \qquad \forall a \in A, i \in a, \tag{2.51}$$

$$U_a^m \geq O_i - \left(1 - I_a^m\right), \qquad \forall a \in \cup_j A_j, i \in a, \tag{2.52}$$

$$\sum_{i \in \mathcal{L}} I_i^L \leq \Gamma^L, \tag{2.53}$$

$$\sum_{a \in A^S} I_a^m \leq \Gamma^{mS}, \tag{2.54}$$

$$\sum_{a \in A^C} I_a^m \leq \Gamma^{mC}, \tag{2.55}$$

$$\sum_{a \in A^S} I_a^b \leq \Gamma^{bS}, \tag{2.56}$$

$$\sum_{a \in A^C} I_a^b \leq \Gamma^{bC}, \tag{2.57}$$

$$\Gamma_{tot} \geq \Gamma^L + \Gamma^{mS} + \Gamma^{mC} + \Gamma^{bS} + \Gamma^{bC}, \tag{2.58}$$

$$\sum_{i \in N} c_i z_i, \leq B, \tag{2.59}$$

$$O, P, U^m, R \geq 0, \tag{2.60}$$

$$I, z \in \{0, 1\}. \tag{2.61}$$

Equations (2.40)–(2.59) are applicability to *any* set of dependency relationships in the framework examined in this chapter, (2.1) and (2.2), where the dependency parameters are not known precisely. However, when dependency relationships or operability values are assumed to be discrete, as is the case with the CJA in (2.17)–

(2.19), this is not advised, both because its discrete nature already makes the analysis robust to unknowns and because the new equations that implement robustness are not easily interpretable.

2.4.3 Defender Formulation: No Robustness

The development of the defender formulation of the optimization, with and without robustness considerations, using the framework of (2.1) and (2.2) is analogous to that of the attacker formulation. There are a number of ways to frame the defender formulation. For illustrative purposes, we assume any leaf node that is not defended is compromised and use w_i as a binary variable to indicate either defense or investment in node i. Equation (2.69) enforces the defense formulation assumption. Constraint (2.70) is the budget that limits the total investment of the defender. Equations (2.63)–(2.69) are analogous to (2.27)–(2.35), although the inclusion of the big M variable and (2.27)–(2.30) can be made simpler because the objective is maximization rather than minimization.

$$\max_{\vec{w}} \Big[O_0 \Big] \tag{2.62}$$

subject to

$$O_j \leq R_j^S, \qquad \forall j \in N : A_j^S \neq \emptyset, \tag{2.63}$$

$$O_j \leq R_j^C, \qquad \forall j \in N : A_j^C \neq \emptyset, \tag{2.64}$$

$$O_j \leq 1, \qquad \forall j \in N : A_j \neq \emptyset, \tag{2.65}$$

$$R_j^S \leq \frac{1}{|A_j^S|} \sum_{i \in A_j^S} P_{(i,j)}^S, \qquad \forall j \in N : A_j^S \neq \emptyset, \left|A_j^S\right| > 0, \tag{2.66}$$

$$R_j^C \leq P_{(i,j)}^C, \qquad \forall j \in N, i \in A_j^C \tag{2.67}$$

$$P_a \leq m_a O_i + b_a, \qquad \forall a \in \cup_j (A_j^S \cup A_J^C), \tag{2.68}$$

$$O_i \leq \ell_i + (G_i - \ell_i) w_i, \qquad \forall i \in N : A_i = \emptyset, \tag{2.69}$$

$$\sum_{i \in N} c_i w_i \leq B, \tag{2.70}$$

$$O, P, R \geq 0, \tag{2.71}$$

$$w_i \in \{0, 1\}, \qquad \forall i \in N. \tag{2.72}$$

2.4.4 Defender Formulation with Robustness

Introducing robustness to the defender formulation can be put into an optimization program by simply supplementing (2.62)–(2.72) with (2.53)–(2.58), and by again noting that, in some cases (2.53)–(2.58) should be supplemented by

$$\sum_{i\in\mathcal{L}} I_i^L + \sum_{a\in A^S}(I_a^m + I_a^b) + \sum_{a\in A^C}(I_a^m + I_a^b) \leq \Gamma_{tot}.$$

$$\min_{\vec{I}^m,\vec{I}^b,\vec{I}^G}\left[\left[\max_{\vec{w}} O_0\right]\right] \tag{2.73}$$

subject to

$$O_j \leq R_j^s, \qquad \forall j \in N : A_j \neq \emptyset, \tag{2.74}$$

$$O_j \leq R_j^c, \qquad \forall j \in N : A_j \neq \emptyset, \tag{2.75}$$

$$O_j \leq 1, \qquad \forall j \in N : A_j \neq \emptyset, \tag{2.76}$$

$$R_j^s \leq \frac{1}{|A_j^S|}\sum_{i\in A_j^S} P_{(i,j)}^s, \qquad \forall j \in N : \left|A_j^S\right| > 0, \tag{2.77}$$

$$R_j^c \leq P_{(i,j)}^c, \qquad \forall i \in A_j^C, j \in N, \tag{2.78}$$

$$P_a \leq m_a O_i + \Delta_a^m U_a^m + b_a + I_a^b \Delta_a^b, \qquad \forall a \in A, \tag{2.79}$$

$$O_i \leq \ell_i + (G_i - \ell_i)w_i - I_i^G \Delta_i^G, \qquad \forall i \in N : A_i = \emptyset, \tag{2.80}$$

$$U_a^m \leq I_a^m, \qquad \forall a \in A, \tag{2.81}$$

$$U_a^m \leq O_i, \qquad \forall a \in A, i \in a, \tag{2.82}$$

$$U_a^m \geq O_i - \left(1 - I_a^m\right), \qquad \forall a \in A, i \in a, \tag{2.83}$$

$$\sum_{i\in\mathcal{L}} I_i^G \leq \Gamma^G, \tag{2.84}$$

$$\sum_{a\in A^S} I_a^m \leq \Gamma^{mS}, \tag{2.85}$$

$$\sum_{a\in A^C} I_a^m \leq \Gamma^{mC}, \tag{2.86}$$

$$\sum_{a\in A^S} I_a^b \leq \Gamma^{bS}, \tag{2.87}$$

$$\sum_{a \in A^C} I_a^b \leq \Gamma^{bC}, \tag{2.88}$$

$$\Gamma_{tot} \geq \Gamma^{mS} + \Gamma^{mC} + \Gamma^{bS} + \Gamma^{bC}, \tag{2.89}$$

$$\sum_{i \in N} c_i w_i \leq B, \tag{2.90}$$

$$O, P, U^m, R \geq 0, \tag{2.91}$$

$$I, z \in \{0, 1\}. \tag{2.92}$$

As was the case for the attacker formulation, (2.73)–(2.92) are applicable to *any* set of dependency relationships in the framework examined in this chapter, (2.1) and (2.2), when the dependency parameters are not known precisely. However, where dependency relationships or operability values are assumed to be discrete, as is the case with the CJA in (2.17)–(2.19), this is not advised both because its discrete nature already makes the analysis robust to unknowns and because the new equations that implement robustness are not easily interpretable.

2.5 Conclusion

This chapter introduced a new simple mission functional dependency representation framework able to encapsulate nearly all practically useful dependency representations as special cases. Its simple formulation, as shown in (2.1) and (2.2), captures exactly or with arbitrarily accurate precision continuous and discrete dependency representations. In Sect. 2.3, FDNA was used to illustrate an example of a continuous dependency representation, SODA was used to illustrate a continuous dependency representation that was arbitrarily accurately approximated, and CJA was used to illustrate a discrete dependency representation. The analysis in this chapter and future analysis based on (2.1) and (2.2) will be applicable to future mission representations of interest. In Sect. 2.4, both an attacker formulation and a defender formulation were presented, each illustrating two scenarios: (1) the scenario where all parameters are known exactly, and (2) the robust scenario where the parameters are only known to be within specified intervals.

References

Deming WE (1951) On statistical techniques in industry as a national resource. In: Bulletin of the international statistical institute 33 (IV) Demography and Labour Statistics: 123–133; quoted on p 125 (Proceedings of the International Statistical Conferences, India 1951)

Garvey P, Pinto C (2009) Introduction to functional dependency network analysis. In: Proceedings from the second international symposium on engineering systems. Massachusetts Institute of Technology, Cambridge. http://esd.mit.edu/symp09/day3.html

Garvey P, Pinto C (2012) Advanced risk analysis in engineering enterprise systems. Chapman-Hall/CRC-Press, Taylor & Francis Group (UK), Boca Raton. ISBN 978-1439826140

Garvey P, Pinto, C, Santos J (2014) Modelling and measuring the operability of interdependent systems and systems of systems: advances in methods and applications. Int J Syst Syst Eng 5(1):1–24

Guariniello C, DeLaurentis, D (2016) Supporting design via the system operational dependency methodology. Res Eng Des 28:53–69

Hastings G, Montella L, Watters J (2009) MITRE Crown Jewel Analysis process, MITRE Corporation MITRE Technical Report MTR090088. 2009

Servi L, Garvey P (2017) Deriving global criticality conditions from local dependencies. Syst Eng 20(4):297–306

Ubagg S (1985) Empirical studies of SODA—a new formula. LHS Proc 31(4):1–5

Watters J, Morrisey S, Bodeau D, Powers S (2009) The risk-to-mission assessment process (RiskMap): a sensitivity analysis and an extension to treat confidentiality issue. The MITRE Corporation. The Institute for Information Infrastructure Protection, Dartmouth, pp 25–26. Pub Rel 09-2994

Zusack S, Guariniello C, DeLaurentis D (2018) Operational dependency analysis of a human mars architecture based on the SODA methodology. In: IEEE Aerospace Conference, pp 1–12

Chapter 3
A Discrete Universal Framework for Mission Dependency Networks Applied to System Identification and Critical Node Analysis

The first step in cooking rabbit stew is catching the rabbit.

Isaac Asimov (1982)

3.1 Introduction

3.1.1 An Overview of Mission Dependency Relationships

The analysis of the complex relationships between an effective mission and the components that contribute to that effectiveness is critical to understanding mission or system of system risks. As systems of systems become more complex and interdependencies more subtle, it becomes imperative to quantify these relationships instead of relying solely on the intuitive judgment of experts familiar with the systems. Once quantified, analysis can be performed to give greater insight into where the greatest risks are, how the risks can be abated via system investments, or how the risks can be leveraged if one wished to impair other's mission.

This chapter presents a general framework for capturing mission dependency relationships that is the discrete parameter counterpart to the continuous parameter framework in Chap. 2. Both frameworks generalize the FDNA representation from Chap. 1. Chapter 3 also presents a method to infer the parameters of this framework from experimental data rather than require a subject matter expert, an SME. In particular, it presents an approach to use simulation experiment output to infer dependency parameters that are most consistent with the experiment. Finally, this chapter presents an algorithm that uses these parameters to identify the critical nodes of the mission.

Many missions correspond to a capability, which has an operability value in [0, 100] that can be measured. A mission dependency representation describes the relationships between an overall mission effectiveness and the components that it depends on to succeed. It is represented as a directed graph where each node represents a resource or an activity, and each directed edge from node i to node j

L. Servi, *Mission Dependency Network Analysis*, Risk, Systems and Decisions,
https://doi.org/10.1007/978-3-032-11260-6_3

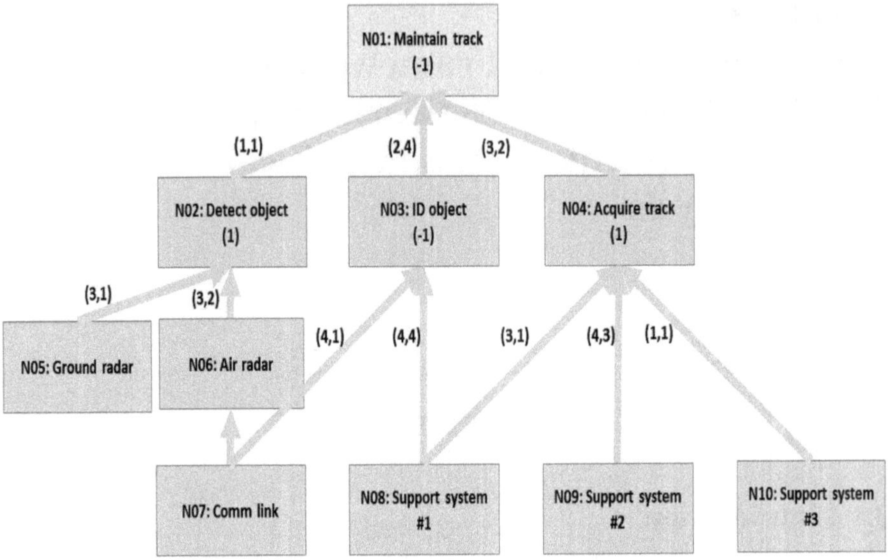

Fig. 3.1 An illustrative example of dependency relationships with FDNA parameters

represents a dependency of parent node j on child node i. These edges often have parameters that specify the nature of the dependency.

The first goal of dependency representation frameworks and related analyses is to determine the impact of degraded or compromised resources/activities on the mission. With a dependency representation, a mission planner can quantify the impact of the impairment of any component or groups of components on the overall mission. This information can be used both defensively when seeking to fortify a mission and offensively to most efficiently degrade a mission.

Consider, for illustrative purposes, the simplified mission dependency representation in Fig. 3.1. Here the key mission is to maintain tracks of identified objects. This can only be done, according to Fig. 3.1, if the object is successfully initially detected, the object is identified, and an initial track is acquired. To detect the object, a ground radar and an air radar are needed. In Fig. 3.1, it is assumed that both the air radar and the capacity to identify an object depend on air/ground communication, but being able to identify an object also depends on "Support system #1." Finally, it is assumed that acquiring a track depends on "Support system #1," "Support system #2," and "Support system #3."

The first steps in analyzing such networks, and in particular in identifying the most critical nodes in such networks, are to (a) map out the topology of the network, as was done in Fig. 3.1, (b) specify clearly defined operability values of the nodes, and (c) clearly define the dependency rules that characterize how the operability of child nodes impacts the operability of their parent nodes. With these components, it is possible to compute how the operability degradation of any node propagates up to the top mission node, "N01- Maintain track." This makes it meaningful to be able to define and seek to identify critical nodes in the network.

This chapter is structured as follows. Section 3.1.2 reviews four representative dependency representations, identifies their common elements, and reviews critical node analysis methods (- more generally, discrete choice optimization methods). This prepares for the introduction of a new general discrete framework in Sect. 3.2, which extracts common elements between the four dependency representations, making each a special case. Section 3.3 introduces an automated method for assigning parameters to a mission dependency representation most consistent with numerous simulation experiments. It can augment or replace the alternative of resource-intensive and time-intensive subject matter expert deliberations that otherwise must occur in order to assign dependency parameters. The approach is validated with numerical results from five experiments that utilize a mission map with 148 nodes and 217 edges. Section 3.4 presents a numerical example of a single mission map as shown in Fig. 3.1, with two different dependency representation assumptions as indicated in Figs. 3.1 and 3.4. In addition, this section describes an algorithm that can identify which set of N nodes in the network that have the greatest impact on the overall mission node under either set of assumptions. Insight gained in this section is useful in understanding potential adversary actions and identifying defensive decisions. Finally, Sect. 3.5 summarizes the contributions of this chapter.

3.1.2 *Technical Background*

This section first reviews a wide spectrum of representative dependency relationships. Each represents a special case of a common discrete dependency relationship framework introduced in Sect. 3.2. Next, a brief review of alternative critical node analysis methods is presented, as well as a brief summary of discrete choice optimization methods; the latter will be used in Sects. 3.3 and 3.4 to identify the child contribution function parameters most consistent with experimental data, as well as to infer the most critical nodes.

3.1.2.1 A Review of Representative Dependency Relationship Frameworks

A number of examples of dependency relationship frameworks exist for characterizing component dependencies, all with different strengths. This section reviews those used by four dependency representations: Crown Jewel Analysis (CJA), Functional Dependency Network Analysis (FDNA), System Operational Dependency Analysis (SODA), and Fault Tree.

CJA employs the RiskMAP, a hierarchical dependency relationship framework that identifies the cyber assets most critical to mission success (Hastings et al. 2009; Watters et al. 2009). It is governed by only three predefined rules that stipulate how a component at one level depends on a component at a level below it. From top to bottom, RiskMAP sequentially maps the overall mission to dependent tasks, dependent information, and finally dependent cyber assets. Since it is easy to understand and able to capture broad dependency relationships, system operators and experts can readily provide inputs reflecting the mission they are aiming to portray. On the other hand, not all complex relationships can be captured with the small set of predefined rules. In its current form in the literature, RiskMAP does not allow for the redundancy of components or dependencies between components at the same hierarchical level, although the latter restriction could be relaxed in a natural way. CJA was also examined in Chap. 2.

FDNA (Garvey and Pinto 2009, 2012; Garvey et al. 2014) employs an alternative dependency relationship framework to RiskMAP, which has a more granular characterization of the relationship between mission components. This comes from defining two continuous-valued parameters for each edge in the network graph. The discretization of these parameters in the context of a Design Structure Matrix (DSM) has also been shown as a useful way to simplify the choices faced by those aiming to map component dependencies (Browning 2001, 2016; Servi and Garvey 2017). With DSM, categories of edges are all given the same parameters. With FDNA, inaccurately identifying these parameters will result in different nodes identified as most critically. This has led to the analysis of robust optimization (Bertsimas and Sim 2004) on FDNA graphs that hedge against the uncertainty associated with assigning point estimate parameters, cf. Chaps. 7 and 8.

The System Operational Dependency Analysis (SODA) methodology is a variation of FDNA constructed by redefining one of the FDNA parameters and adding two more parameters (Guariniello and DeLaurentis 2017). This dependency representation allows more fine-tuning and more accurate representation of component relationships. However, setting these parameters can be an obstacle for mission modeling, because it takes more effort to properly assign all four of its parameters. A strict adherence of the SODA methodology contains nonlinearity, which may impact the performance of some algorithms.

Finally, the Fault Tree methodology is used in reliability analysis to identify the sub-events and factors that contribute to a catastrophic event or failure (Reliability Information Analysis Center 2005). Its dependency relationship framework is binary, as an event (modeled as a mission component) can either occur or not, and uses simple AND and OR nodes to capture the relationships between events. While this dependency representation is easy to understand and use, it does not allow for more subtle relationships to be captured.

A set of input components is common for many dependency representations: a node metric, child contribution functions, and rules for aggregating the child contributions. These components are explained below, followed by Table 3.1, which shows how they are applied to the aforementioned four dependency representations.

Node Metric To measure the effectiveness of the mission and its dependencies, there must be some common metric to assess the status of the nodes. This may be an operability value, a risk level, or some other metrics of the state of the node. This node metric is a scalar value of the node and may be able to take on continuous or discrete values. The value of the node metric of a parent node can be calculated from the values of its children's node metrics using the children contribution functions. Therefore the node metric values of the leaf nodes, the bottom nodes in the map, ultimately propagate through the mission dependency representation to the mission node.

Child Contribution Functions The node metric value of a parent node depends on the metric values of its children. The exact relationships are defined by child contribution functions (CCFs). There are two types of child contribution functions, single CCFs and multi-CCFs as illustrated in Table 3.1. Single CCFs are functions of a single dependent child node metric value, whereas multi-CCFs are functions of two or more children of the parent. For example, in FDNA, the Criticality of Dependency (COD) function is a single CCF, and the Strength of Dependency (SOD) function is a multi-CCF.

MaxMin Rules A parent might have more than one contribution from a combination of single CCFs and multi-CCFs. Therefore, the dependency representation must have rules for aggregating these multiple contributions. While in principle any valid mathematical or logical combination of contributions could be used, for the purposes of this chapter, a minimalist approach is adopted. Such an approach decreases the danger of System Identification algorithms, discussed in Sect. 3.3, overfitting the data. In this chapter, we assume that the parent either performs as poorly as the worst contribution or as well as the best contribution. These two possibilities correspond to taking the minimum or maximum contribution and must be specified for each node with children.

Table 3.1 illustrates how the components of RiskMAP, FDNA, SODA, and Fault Trees fit within these categories using the variable ℓ_i as the node metric for node i. ℓ_i takes on discrete values for the RiskMAP model in $\{0, 35, 70, 90\}$ and for the Fault Tree model in $\{0, 1\}$. For the FDNA and SODA model, ℓ_i has continuous values in $[0, 100]$. RiskMAP and Fault Tree assess risk and failures in these dependency representations, whereas FDNA and SODA are assessing operability. Therefore, a "high" node metric is favorable for FDNA and SODA but is unfavorable for RiskMAP and Fault Tree. Thus, the maximizing and minimizing aggregations have opposite effects for these two sets of dependency models. Due to the discrete nature of the RiskMAP model, the value of ℓ_i has a relative but not an absolute interpretation.

Table 3.1 presents a synthesis of the four dependency representations into a common, generalized framework. In particular, the parameter a_{ij} is a threshold parameter used in RiskMAP with the parameter b_{ij}. RiskMAP classically does not present the dependency rules with piecewise functions but is equivalent to the rules in Table 3.1. Specifically, the parameters a_{ij} and b_{ij} can be set according to the three dependency rules in Watters et al. (2009). FDNA has two parameters per edge: the Strength of Dependency (SOD), α_{ij}, and the Criticality of Dependency

Table 3.1 Comparison of RiskMAP, FDNA, SODA, and Fault Tree representations

Dependency representation	Node metric	Single child contributions	Multi-child contributions	Contribution aggregation rules
RiskMAP	Risk score: discrete $\ell_i \in \{0, 35, 70, 90\}$	$\ell_j = \begin{cases} 0 & \ell_i < a_{ij} \\ b_{ij} & \ell_i \geq a_{ij} \end{cases}$	NA	Maximum
FDNA	Operability: continuous $\ell_i \in [0, 100]$	$\ell_j = \ell_i + \beta_{ij}$	$\frac{1}{\lvert C(j) \rvert} \sum_{i \in C(j)} (\ell_i \alpha_{ij} + 100(1 - \alpha_{ij}))$	Minimum/maximum
SODA	Operability: continuous $\ell_i \in [0, 100]$	$\ell_j = \frac{(100)}{\gamma_{ij}} \ell_i + W_i^{\lambda}(100 - \beta_{ij})$	$\frac{1}{\lvert C(j) \rvert} \sum_{i \in C(j)} (\ell_i \alpha_{ij} + SE_j(1 - \alpha_{ij}))$	Minimum/maximum
Fault Trees	Failure state: Boolean $\ell_i \in \{0, 1\}$	$\ell_j = \ell_i$	NA	Minimum/maximum

(COD), β_{ij}, that are used in calculating the operability of the parent nodes. The details of FDNA are described in Garvey and Pinto (2009, 2012); Garvey et al. (2014). SODA is a variation of FDNA, so it also specifies α_{ij} and β_{ij} for each edge (with the slight difference that $\beta_{SODA} = 100 - \beta_{FDNA}$) as well as needing three additional parameters: γ_{ij} (the impact of the dependency for each edge), SE_j (a "self-effectiveness" term), and λ, which is typically set to 0.1. SODA also uses the variable W_i to be the average operability value of all of the dependent children of node j, except i assuming j has more than one dependent node. If node j has exactly one dependent node, then W_i is set to 1. The details of SODA are described in Guariniello and DeLaurentis (2017). For FDNA and SODA, $C(j)$ denotes the dependent children of node j, i.e., the nodes whose operability levels directly impact the operability level of j. The default node MaxMin rule in FDNA and SODA is minimum, but the literature also defines "OR nodes," which are logical OR gates and correspond to taking the maximum (Servi and Garvey 2017). Fault Trees contain AND/OR logic nodes, so a maximum MaxMin rule represents an OR gate, and a minimum rule represents an AND gate.

The general dependency representation framework introduced in this chapter is flexible and intuitive. In a very general context, it allows the user to specify as many discrete node operability values and as many dependency rules as is appropriate for their system. It captures the granularity of FDNA and SODA, while also being able to capture the smaller data requirements of representations such as RiskMAP and Fault Trees. This generalized framework represents these four dependency representations as illustrative special cases. The discrete nature of this general dependency representation framework creates the opportunity to use experimental output to infer which dependency rules should be assigned to each edge in a manner most consistent with the experiments, as well as being amenable to algorithms that identify the critical nodes in the network.

3.1.2.2 A Review of Critical Node Analysis Methods

As noted in Chap. 1, the criticality of a node is a concept with many reasonable definitions. It is not possible to compare the relative effectiveness of each alternative approach, as each defines criticality differently, and each computes criticality according to that definition. To effectively engineer systems, it is important to have a large toolbox of critical measures and approaches.

The vast literature on critical node analysis can be grouped into two polarized viewpoints. One view is that the network topology is sufficient for critical node identification. Examples of such a viewpoint include (Ahuja et al. 1988; Wasserman et al. 1994) that examine the Birnbaum Structural Importance (BSI) (Birnbaum 1969) (and Chap. 5) in a Fault Tree (Kuo and Zhu 2012), the degree of centrality of the nodes (Wasserman et al. 1994), the global clustering coefficients (Costa et al. 2007), the cutsets (Ahuja et al. 1988), the degree of the nodes (Albert et al. 2000), the pairwise connectivity (Shen et al. 2013; Ventresca and Aleman 2015), and the shortest paths to a key node (Albert et al. 2004). With this viewpoint, the impact of the loss of nodes on system homogeneity, diffusion speed, and diffusion (Chen 2016) uses solely the topology to identify critical nodes.

An alternative view is that the topology alone is not enough information, since nodes and edges are quite heterogeneous, and the differences between them have important bearing on which nodes are the most critical. This viewpoint requires vastly more information before an analysis can be performed, but the resulting analysis reflects more nuance. Examples of such a viewpoint include FDNA (Garvey and Pinto 2009, 2012; Garvey et al. 2014) and SODA (Guariniello and DeLaurentis 2017). The most appropriate viewpoint depends on the specific application of interest.

This chapter adopts a middle-ground viewpoint by presenting a flexible analytical framework in Sect. 3.2, which permits the user to specify the level of granularity of nodal and edge information needed. Hence, in Sect. 3.2, special-case examples of the framework include a Fault Tree analysis and RiskMAP, which are indicative of the first viewpoint, as well as a discretized version of FDNA and SODA, which are indicative of the second viewpoint.

3.1.2.3 A Review of Some Discrete Choice Optimization Methods

Discrete choice optimization methods solve the problem of selecting the best subset of elements from a finite discrete set using a specified objective function to evaluate each choice. Two such algorithms are presented in this chapter. In Sect. 3.3, a genetic algorithm is presented to identify the single child and multi-child contribution aggregation rules most consistent with experimental data. In Sect. 3.4, a greedy algorithm is presented to find a set of N nodes that impact the mission more than any other set of N nodes.

Genetic algorithms (Anderson-Cook 2005; Davis 1991; Mitchell 1996) are an example of a bioinspired optimization algorithm (Binith and Sathya 2012), which does not become trapped in a local optimal point. There are four basic steps of a genetic algorithm: (i) initiation, (ii) selection, (iii) crossover, and (iv) mutation. The initiation step consists of constructing an initial set of candidate solutions. Next, the selection step entails evaluating the current candidate set of solutions to guide the elimination of those less than a threshold and setting aside candidates in pairs for those above the threshold. In the crossover step, components of each pair of candidate solutions are randomly selected to be exchanged to create a similar but different set of solutions. Finally, the mutation step consists of randomly altering some components of the candidate solutions. The selection, crossover, and mutation steps are repeated multiple times during the execution of the genetic algorithm.

In contrast, a greedy algorithm iteratively adds additional elements to a set of what has already been selected and in a manner that is always the best choice with respect to the objective function. This is an example of a local optimization method that will find a local minimum and tends to be very fast but provides no guarantee of global optimality. The algorithm will be less effective if the selection of one element impacts the relative importance of other elements in a non-separable manner. A related local optimization approach begins with a tentative solution and iteratively replaces a small group of elements already selected with elements not yet selected with the property that the result of the exchange improves the objective value.

The genetic and greedy algorithms in this chapter seek to balance computation time with effectiveness in solving discrete choice optimization problems of interest. There is vast literature of both variations of the algorithms presented in this chapter (Goldberg and Deb 1991) and alternative algorithms (Boussaid et al. 2013). Of particular note is a line of research using genetic algorithms to identify clusters of nodes that have strong interdependencies (Yu and Goldberg 2004). Alternatives include the tabu search (Glover 1986), bioinspired optimization algorithms (Binith and Sathya 2012) such as ant colony optimization (Dorigo and Blum 2005) or bee colonies (Walker et al. 1993), and mixed linear integer programming (MLIP) (Schrijver 1998). For each method, it is important to ensure that nonlinear relationships between the optimization parameters are sufficiently acknowledged when important. For example, in failure analysis it is sometimes insufficient to find only first-order effects and that second- and third-order effects (i.e., a combination of two or three independent effects) should be searched for Grindal et al. (2005); Kuhn et al. (2004, 2008).

3.2 A Generalized, Discrete Dependency Representation Framework

This section presents a generalized, discrete dependency representation framework that captures the essence of the four models in Table 3.1. It permits a modeler to tune the generalized framework to one of the four models or to their specific needs.

The dependency representations must be acyclic as it is not meaningful for dependencies to be cyclic. For example, it is not meaningful for node i to depend

Table 3.2 Capturing nature of RiskMAP, FDNA, and SODA dependency relationships in a general framework

Dependency representation	Set of allowable node values, $\mathcal{L}^*$	Single child contr. rules, **B**	Multi-child aggr. contr. rules, $\vec{a}$	MaxMin rules, m_j
RiskMAP	{0,35,70,90}	$\begin{bmatrix} 0 & 0 & 3 & 3 \\ 0 & 0 & 2 & 2 \\ 0 & 0 & 1 & 1 \end{bmatrix}$	0	$1\ \forall j$
FDNA	{0,25,50,75,100}	$\begin{bmatrix} 0 & 1 & 2 & 3 & 4 \\ 1 & 2 & 3 & 4 & 4 \\ 2 & 3 & 4 & 4 & 4 \\ 3 & 4 & 4 & 4 & 4 \end{bmatrix}$	[0, 1, 2, 3, 4]	$1\ \forall j \in$ OR nodes $-1\ \forall j \notin$ OR nodes
SODA	{0,25,50,75,100}	$\begin{bmatrix} 0 & 1 & 2 & 3 & 4 \\ 0 & 2 & 4 & 4 & 4 \\ 1 & 3 & 4 & 4 & 4 \\ 2 & 3 & 4 & 4 & 4 \\ 2 & 4 & 4 & 4 & 4 \end{bmatrix}$	[0, 1, 2, 3, 4]	$1\ \forall j \in$ OR nodes $-1\ \forall j \notin$ OR nodes
Fault Trees	{01,1}	[0 1]	0	$1\ \forall j \in$ OR nodes $-1\ \forall j \in$ AND nodes

on node j *and* for node j to depend on node i. However, it may be the case, for example, that part of node i depends on part of node j, and a different part of node j depends on a different part of node i. In this case, i and j could each be split into two separate nodes to create an acyclic mapping, which is meaningful. More generally, the splitting of nodes into parts is a useful practical approach to creating acyclic mission dependency representations (Pinto and Garvey 2012).

Section 3.2.1 describes general considerations in the design of a dependency representation, followed by Sect. 3.2.2, which introduces the notation. In Sect. 3.2.3, the four dependency representations are put into the generalized discrete model and summarized in Table 3.2.

3.2.1 Design Choices

The first design choice is for the generalized dependency representation framework to only allow a node metric to take on discrete values. The nodes in a mission network can represent physical items such as hardware or less tangible items such as a task that needs to be performed. It is difficult to precisely calibrate these different node types to the same metric scale when translating a real system and mission into a network representation. Instead of defining a continuous metric scale, the use of discrete levels can more accurately capture the known level of precision and reduce the algorithm's computation complexity.

The second design choice deals with the multi-child contributions in the generalized framework. In FDNA and SODA, the multi-child contributions are essentially weighted averages of the children's operability values, where nodes with a greater importance have a higher weight. Taking this idea into a discrete context, the multi-child contribution in the generalized framework is a weighted average of the child node operability values rounded to the nearest discrete metric value.

3.2.2 *Notation*

Below is the notation used throughout this chapter to characterize the dependency representations, assuming the node metric value of node j is computed from the single child contributions of its dependent children and a possible multi-child contribution.

- $C(j)$ is the set of dependent children nodes of node j. This can be very general, e.g., two nodes that are children of node j might have the property that one is also a dependent child of the other.
- $\mathcal{L} = \{0, 1, 2, \ldots, |\mathcal{L}| - 1\}$ is a finite set of discrete *indices* of the node metric values.
- $I_i \in \mathcal{L}$ is the node metric index of node i.
- $\mathcal{L}^* = \{\ell_0, \ell_1, \ell_2, \ldots, \ell_{|\mathcal{L}|-1}\}$ is a finite set of node metric values. The node metric values might, for example, be operability levels or risk levels. The values do not have to correspond to equidistant levels. For example, $\mathcal{L}^*$ might equal $\{0, 30, 80, 100\}$. By construction, $|\mathcal{L}| = |\mathcal{L}^*|$.
- $\ell_I \in \mathcal{L}^*$ is the node metric value of a node with a node metric index of $I \in \mathcal{L}$.
- $\mathcal{R}_p = \{1, 2, \ldots, |\mathcal{R}_p|\}$ is a finite set of discrete *indices* of the rules for the possible single child contribution functions and multi-child contribution functions for $p = 1$ and $p = 2$, respectively.
- $r_p(i, j) \in \mathcal{R}_p$ is the index of the rule for the single child contribution and multi-child contribution functions, for $p = 1$ and $p = 2$, respectively, relating node j to its dependent child node i. In addition, $r_p(0, j)$ is an additional rule index associated with parent node j which is typically used for $p = 2$.
  The precise details of the child contribution functions are specified by the matrix **B**, the vector $\vec{a}$, C_{0j}, and the $C_{i,j}$'s defined in (3.1) and (3.2) below.
- m_j is the MaxMin rule assigned to each parent node j. It is 1 if node j maximizes all contributions and -1 if it minimizes all contributions.
- **B** is an $|\mathcal{R}_1| \times |\mathcal{L}^*|$ matrix of elements of node metric values. The rows of **B** refer to the indices of the applicable rule for a single child contribution, and the columns refer to the metric value of the child node. Specifically, if the single child

contribution rule between node i and node j is set to $r_1(i, j) = k \in \mathcal{R}_1$ and node i is at node metric value $\ell_{I_i} = \ell_{\mathcal{I}}$, the single child contribution is $\mathbf{B}_{k,\mathcal{I}} \in \mathcal{L}^*$.

- $\vec{a}$ is a vector of $|\mathcal{R}_2| + 1$ dimensions that determines the multi-child aggregate contribution of the child node i to parent node j based on the rule index of edge (i, j), $r_2(i, j)$. The multi-child contribution rule from all the children of node j is given by (3.1). The $(|\mathcal{R}_2| + 1)$th component of $\vec{a}$ corresponds to the rule index, $r_2(0, j)$, associated only with the parent node j.

The following notations are related to calculating the node metric values:

- C_{0j} is the multi-child contribution of the children nodes $i \in C(j)$ of node j. It captures the importance of the (i,j) link and the level, ℓ_{I_i}, of node i. This is given by

$$C_{0j} = a_{r_2(0,j)} + \sum_{i \in C(j)} a_{r_2(i,j)} \ell_{I_i}, \tag{3.1}$$

rounded to the nearest value of $\mathcal{L}^*$.

- C_{ij} is the single child contribution from node i to node j. This is given by

$$C_{ij} = \mathbf{B}_{r_1(i,j),I_i} \in \mathcal{L}^* \qquad \forall\, i \in C(j). \tag{3.2}$$

Therefore, the single child contribution from node i to j is a mapping from the node metric value ℓ_{I_i} to a node metric value in $\mathcal{L}^*$ determined by the rule $r_1(i, j)$.

- The node metric value of node j as a function of the operability values of its children nodes, $i \in C(j)$, is given by

$$\ell_{I_j} = \begin{cases} \max_{\{i \in C(j) \cup \{0\}\}} C_{ij} & \text{if } m_j = 1, \\ \min_{\{i \in C(j) \cup \{0\}\}} C_{ij} & \text{if } m_j = -1. \end{cases} \tag{3.3}$$

For the special case where there is no multi-child contribution for node j, the level of node j is given by

$$\ell_{I_j} = \begin{cases} \max_{\{i \in C(j)\}} C_{ij} & \text{if } m_j = 1, \\ \min_{\{i \in C(j)\}} C_{ij} & \text{if } m_j = -1. \end{cases} \tag{3.4}$$

In summary, depending on the value of m_j, the node metric value of node j is the maximum or minimum of the single child contributions of its dependent children, $i \in C(j)$, and a possible multi-child contribution. The single child contribution of dependent child i is a node metric value based on the node metric index of node i and the specified single child contribution rule between node i and node j. Recall that the node metric index of node i is $I_i \in \mathcal{L}$ and the node metric value of node i equals ℓ_{I_i}. Hence, the single child contribution rule between node i and node j is rule index $r_1(i, j) \in \mathcal{R}_1$, and the single child contribution of node i on node j has a node metric value of $B_{r_1(i,j),I_i}$. The multi-child contribution to node j, if there is one, is based on the metric values of two or more of the dependent child nodes of j, i.e., nodes i for $i \in C(j)$, and the specified multi-child contribution rule between each i and j. Hence, the multi-child contribution rule between each $i \in C(j)$ and node j is $r_2(i, j)$, and the multi-child contribution to j has a metric value given by $C_{0j} = a_{r_2(0,j)} + \sum_{i \in C(j)} a_{r_2(i,j)} \ell_i$, rounded to the nearest value of $\mathcal{L}^*$.

3.2.3 *Capturing the Nature of Four Different Dependency Representations*

Using the generalized framework defined in the previous section, we now demonstrate how it can capture the characteristics of the four dependency representations in Table 3.1. The generalized framework parameter values that represent each dependency representation are presented in Table 3.2.

RiskMAP

In RiskMAP (Watters et al. 2009), nodes are at one of the four levels: *operational*, *work-around*, *degradation*, and *failure*. We can represent these nodal capabilities as $\mathcal{L} = \{0, 1, 2, 3\}$, respectively, in the generalized framework. The set of rules in RiskMAP, which could be applied to different child nodes, is:

1. Rule 1: A failure or degradation of a child node results in a failure of the parent node; otherwise, the parent is operational.
2. Rule 2: A failure or degradation of a child node results in a degradation of the parent node; otherwise, the parent is operational.
3. Rule 3: A failure or degradation of a child node results in a work-around status of the parent node; otherwise, the parent is operational.

The RiskMAP rules are designed to capture the observation that the risk level of some child node impacts its parent node differently than others. These are all single child contributions, so $\mathcal{R}_1 = \{1, 2, 3\}$. Rule 1 is encoded in the first row of **B** as $B_{1,3} = 3$, $B_{1,2} = 3$, $B_{1,1} = 0$, and $B_{1,0} = 0$. The second rule can be encoded as $B_{2,3} = 2$, $B_{2,2} = 2$, $B_{2,1} = 0$, and $B_{2,0} = 0$. The third rule can be encoded as $B_{3,3} = 1$, $B_{3,2} = 1$, $B_{3,1} = 0$, and $B_{3,0} = 0$. Hence, the RiskMAP rules can be encoded in matrix **B** as shown in Table 3.2. There are no multi-child contributions in RiskMAP, so $\vec{a} = 0$ resulting in (3.4) and not (3.3) applying. In RiskMAP, the level of node j is the *maximization* of the level (risk) of its children nodes $C(j)$, so $m_j = 1$ for all j.

Hence, RiskMAP is simply

$$\ell_{I_j} = \max_{\{i \in C(j)\}} [C_{ij}] = \max_{\{i \in C(j)\}} [\mathbf{B}_{r_1(i,j),I_i}]. \tag{3.5}$$

FDNA

FDNA uses a continuous node metric value corresponding to the node's operability level that can take on values between 0 and 100. Because nodes in the generalized framework in this chapter only take on discrete levels of the node metric, FDNA cannot be represented exactly. However, parameters for the generalized framework can be chosen to create a discretized version of FDNA. For example, $\mathcal{L} = \{0, 1, 2, 3, 4\}$ might correspond to operability values $\mathcal{L}^* = \{0, 25, 50, 75, 100\}$ or $\mathcal{L}^* = \{0, 10, 50, 90, 100\}$.

In terms of the single child contribution rules in FDNA, the β parameter can take on any value in [0,100]. Instead of accommodating the infinite number of possible rules, we choose a representative subset of rules shown in the **B** matrix in the second row of Table 3.2 with $\mathcal{R}_1 = \{1, 2, 3, 4\}$. The four rules in **B** can be thought of as FDNA Criticality of Dependency rules with β=0, 25, 50, 75 for each respective row, where ℓ_{I_i} = 0, 25, 50, 75, 100 for I_i = 0, 1, 2, 3, and 4, respectively.

From Table 3.1 and (3.1), the correspondence between the multi-child contribution in FDNA and this generalized framework is

$$a_{r_2(i,j)} = \begin{cases} \frac{1}{|C(j)|} \sum_{i \in C(j)} (100(1 - \alpha_{i,j})) & \text{if } i = 0, \\ \frac{\alpha_{i,j}}{|C(j)|} & \text{if } i \neq 0. \end{cases} \tag{3.6}$$

As noted in Servi and Garvey (2017) and Chap. 1, some nodes may be maximizing OR nodes, while the rest of the nodes use a minimum aggregation. This can be directly represented in the generalized framework by setting m_j to 1 for the OR nodes and -1 for the remaining nodes.

SODA

Because SODA is built on FDNA, everything stated about capturing FDNA-like dependency in the generalized framework holds for SODA. SODA also includes two more parameters in the single child contribution functions (known as Criticality of Dependency in SODA and FDNA), which allow for more fine-tuning of the dependency relationship. There is a nonlinear term $(W_{i,j})^\lambda$, where $W_{i,j}$ is the average of the operability values of all the dependent children of node j except j, which also needs to be approximated discretely in a manner analogous to (3.6) and the parameter, λ, is typically set to 0.1. Table 3.2 shows an example of creating a SODA-like network with the generalized framework assuming $\lambda = 0$. Again, the rules in **B** aim to be representative of the criticality of dependency formulas, as it is impossible to capture the infinitely many rules generated by these continuous parameters.

Fault Trees

Fault Trees are binary in that every node is either in an operable state or in a failure state. This is represented by $\mathcal{L} = \{0,1\}$, where 0 is the operable state, and 1 is the failure state. Any Fault Tree with gates that are combinations of both ORs and ANDs can also be represented by a larger one, where each node is either an OR node or an AND node. Hence, without loss of generality, it will be assumed that this is the case. The rule for node-level propagating effects from children to parent nodes is a simple single child contributing rule: If a child is in a failure state, the contribution is a failure state; if the child is in an operable state, the contribution is an operable state. Therefore, the **B** matrix is simply a single row where the output contribution is the state of the child, and the MaxMin rule for a node is $m_j = 1$ in the case of OR node (maximizing nodes) and $m_j = -1$ for the cases of AND node (minimizing nodes). There is no need for a multi-child contribution aggregation rule, so $\vec{a}$ is set to 0.

3.3 Identification of System Dependency Parameters

Part of the impetus for creating a generalized dependency representation framework with discrete values and a discrete number of rules is the difficulty and time-consuming nature in assigning parameters to edges within a functional mission network. Reducing the number of flexible rules allows the user to assign dependency relationships that are relevant for their system without having to decide between too many options. Frequently, the robustness of the system insulates this identification challenge from being unduly impaired. However, even with this more streamlined approach to dependency representation, the assignment of rules may be too burdensome for the mission planners. This is particularly true for large missions and mission planners who are not intimately involved with every aspect of mission dependencies. A related challenge occurs in the area of Control Theory, where there is a need to develop a model identification procedure, using what is referred to as classical "System Identification" (Ljung 1998). A more subtle version of this challenge shifts the emphasis to "Identification for Control" (I4C) (Gevers 2005), where the focus is not on obtaining the best possible model that fits the data, but to obtain a model fitting well enough to permit desired control performance.

In a similar vein, this section develops a System Identification algorithm of the mission dependency map with the goal of automatically identifying parameters that can be used for analysis of the mission in general and, as will be discussed in Sect. 3.4, identification of critical nodes in particular. This complements what can be achieved using a subject matter expert (SME) approach to parameter identification and creates a set of parameters that could be used as a starting point by experts to fine-tune parameters for related but similar missions. Figure 3.2 reflects the broader goal of supplementing an SME with simulation and a genetic algorithm.

In this section, a new algorithm is first presented to infer the dependency relationship parameters from simulation experiments and is then tested under five different experiments on a network consisting of 148 nodes and 217 edges. Both the accuracy of the results and the speed of the algorithm are reported.

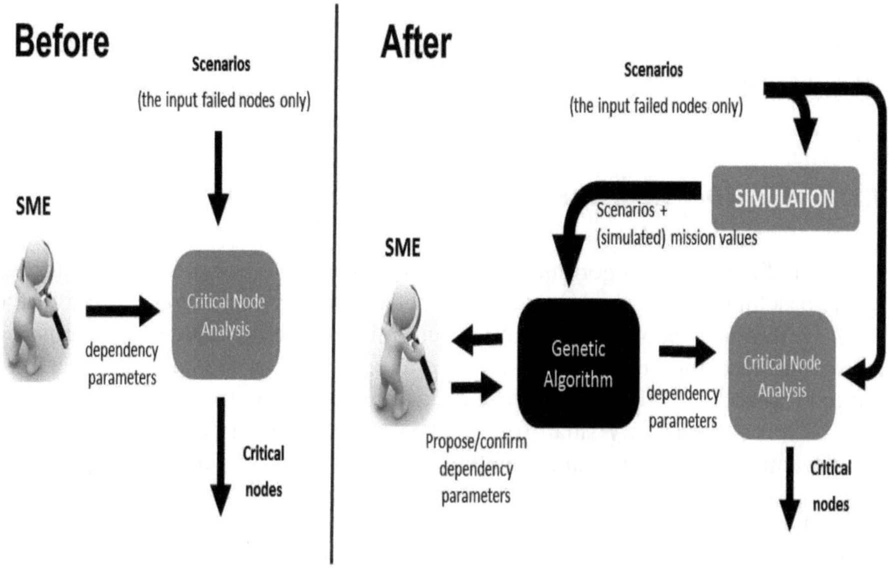

Fig. 3.2 A proposed process for identifying dependency relationship parameters

3.3.1 System Identification Inputs

To identify the edge parameters that define the dependencies throughout the graph, some prior knowledge about the mission mapping must be known. The inputs for the System Identification algorithm are:

- **Data**: Data for a set of scenarios that provides information about the overall mission status given certain nodes or groups of nodes is in a failure state. Each scenario consists of failed nodes as input and a mission value as output. This scenario data is split up into a training set, on which the algorithm is performed, and a test set, with which the results are assessed. This can be generated from a simulation of the system (if available) or from subject matter expertise.
- **Search Space**: A mission mapping includes the nodes, the aggregation rules for each node (i.e., the value of each m_j), and the topology of each parent-child dependency edge between nodes (without the numerical value of the edge rules specified). This will be referred to as the baseline graph. This might be motivated by a related or previous mission that is better understood. The search space is then the set of possible edge rules with the numerical values specified. The simplest dependency representations will have Fault Tree-like rules where failures cascade. More complex missions will have rules that incorporate partial degradation exemplified by other models listed in Table 3.2. These can be chosen based on an SME's understanding of the mission and its complexity. To complement or replace the use of SMEs, an algorithm such as the one below could be used to infer them.

To validate the genetic algorithm, many scenarios must be generated assuming the parameters defining true edge rules are known, which is not in the training set. For experimentation purposes, the true edge rules might be constructed randomly.

3.3.2 *Genetic Algorithm for System Identification*

The goal of the genetic algorithm is to find the edge rules associated with each edge of the baseline graph, which have the property that they minimize the sum of the square differences between the ground-truth mission metric value and the mission metric values computed from these edge rules. It is often the case that there are multiple associations that can produce behavior that mirrors the input scenario data with an error that is very small (or equal to zero). This suggests that choosing a simple genetic search algorithm from the vast of literature of genetic algorithm, cf., Davis (1991); Anderson-Cook (2005); Mitchell (1996), will suffice.

As illustrated in Fig. 3.3 and conveyed in more detail in Algorithm 1, the genetic algorithm starts with a set of randomly generated edge rules for each edge. These will be referred to as candidate assignments, and when a candidate assignment is applied to the baseline graph, this new graph is called the candidate graph. Each candidate graph is assessed against the scenarios in the training set. Specifically, for each scenario, the input failure nodes are set in the candidate graph, and the mission node metric value is calculated using the candidate assignments of edge rules. The difference between this and the "ground-truth" assignment (using the assumed true edge rules) is squared and added to the total error for that candidate assignment. This

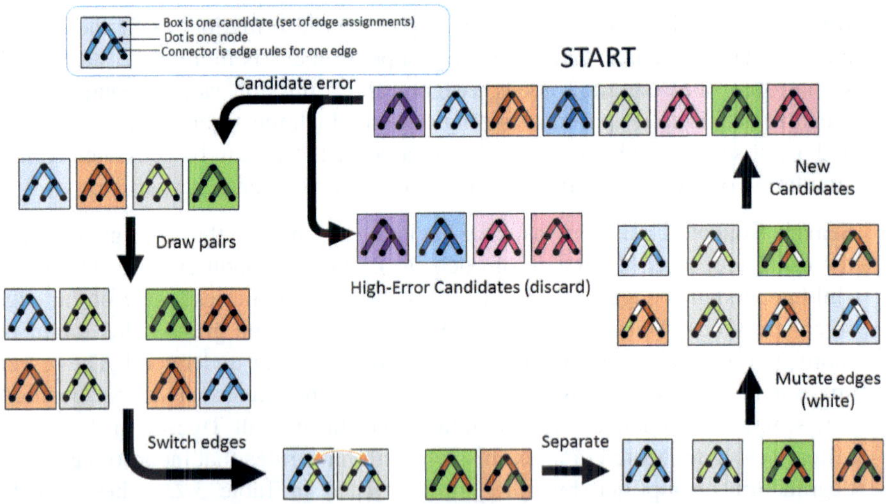

Fig. 3.3 The genetic algorithm

total square error determines the "fitness" of the candidate graph, where a low error corresponds to a high fitness. Candidates with fitness higher than a defined threshold value are selected for the next stage of the algorithm. In this next stage, pairs of candidate assignments are drawn with replacement from the selected set, and the pairs randomly switch each corresponding edge rule with a specified probability. Once the "offspring" of these selected candidates are created, mutations within each offspring edge assignment can occur with a different, prespecified probability. The mutations swap out the edge rule specified with a randomly selected rule. Finally, the mutated offspring become the new set of candidate edge assignments, and this process iterates continuously until an edge assignment with zero error for all scenarios is created or the maximum number of iterations is reached.

Algorithm 1, as presented in the Appendix, provides the pseudo-code for this approach. This algorithm references a function CALCMISSIONNODE(`network`, `scenarioInputs`), which takes as input a graph with edge rules assigned and a set of nodes to set failure mode and outputs the mission node's metric value.

3.3.3 Testing the System Identification Algorithm

To test the System Identification algorithm, five ground-truth dependency relationship parameters were generated that used a graph of 148 nodes and 217 edges. Five numerical experiments were created from the five sets of node metric values and rules shown in Table 3.3. For each edge, ground-truth rules were created by assigning a random vector value to $\vec{a}$ and random index values to $\mathbf{B}$. With the five ground-truth representations, input datasets used by the genetic algorithm could be created to document the mission metric value, i.e., the metric value of the mission node given that random nodes or groups of nodes were considered out of service or at the lowest node level. These provided the `scenarioMissionValues` referenced in the algorithm that are used to calculate the error associated with a particular candidate solution.

The dataset for each mission map was split into training and test sets, stratified by the impact of the scenario on the mission node level, with the training set approximately four times larger than the test set. For example, the dataset generated from the network with the rules from Test Case 1 of Table 3.3 was split into a training and a test set that had the same proportion of scenarios that drove the mission node level to 0 and the same proportion of scenarios that left the mission node level at 1. The genetic algorithm used the training set as an input and four parameters were set: the number of candidates per generation, the crossover probability, the mutation probability, and the maximum number of iterations. The number of candidates per generation was set to 50. The crossover probability, the probability that two candidate edge sets will exchange an edge assignment, was set to 0.35. The mutation probability, the probability that an edge assignment within a candidate edge, was set to 0.1. The maximum number of iterations was set to 200. These parameters were the same for the testing of each of the five different ground-

Table 3.3 Metric values and rules defining five test mission representations for system identification algorithm

Experiment graph	Set of allowable node metric values, $\mathcal{L}$	Single child contr. rules, **B**	Multi-child aggr. contr. rules, $\vec{a}$
1	[0,1]	$\begin{bmatrix} 0 & 1 \\ 1 & 1 \\ 0 & 0 \end{bmatrix}$	[0,1,2]
2	[0,1,2]	$\begin{bmatrix} 0 & 0 & 1 \\ 0 & 1 & 2 \\ 1 & 1 & 2 \\ 1 & 1 & 2 \\ 0 & 0 & 2 \\ 0 & 1 & 1 \end{bmatrix}$	[0, 2, 5]
3	[0,1,2,3]	$\begin{bmatrix} 0 & 1 & 2 & 3 \\ 1 & 1 & 2 & 3 \\ 0 & 0 & 0 & 3 \\ 0 & 0 & 1 & 2 \\ 0 & 0 & 2 & 2 \end{bmatrix}$	[0, 2, 4, 8]
4	[0,1,2,3,4]	$\begin{bmatrix} 0 & 1 & 2 & 3 & 4 \\ 1 & 1 & 2 & 3 & 3 \\ 0 & 0 & 0 & 2 & 4 \\ 0 & 1 & 1 & 2 & 4 \\ 0 & 0 & 2 & 3 & 4 \\ 1 & 1 & 2 & 2 & 3 \end{bmatrix}$	[0,1,2,3]
5	[0,1,2,3,4,5]	$\begin{bmatrix} 0 & 1 & 2 & 3 & 4 & 5 \\ 0 & 0 & 1 & 2 & 4 & 5 \\ 1 & 2 & 4 & 4 & 4 & 5 \\ 0 & 0 & 0 & 1 & 3 & 4 \\ 0 & 3 & 4 & 4 & 4 & 5 \\ 0 & 0 & 0 & 2 & 2 & 3 \\ 0 & 1 & 1 & 1 & 3 & 4 \\ 0 & 0 & 0 & 0 & 0 & 5 \\ 0 & 0 & 0 & 1 & 1 & 2 \\ 1 & 2 & 3 & 4 & 5 & 5 \end{bmatrix}$	[0,1,2,3,5,8,13]

truth graphs. Table 3.4 shows the results of the test cases, where each row represents average results from three trials of the genetic algorithm. The first column references the metric values and rules from Table 3.3 that were used to create the ground-truth network and were used in the genetic algorithm. The second column provides the average number of iterations performed before the algorithm was terminated. Note an entry of 200 in this column means the algorithm reached the maximum in all three trials. The third column provides the average time in minutes that the algorithm took before it was terminated. The fourth and fifth columns provide the average error rate

Table 3.4 System identification algorithm results

Experiment graph	Average iterations	Average time (minutes)	Training set error rate	Test set error rate
1	200	57.2	0.021	0.020
2	200	63.4	0.058	0.204
3	89	21.4	0.000	0.012
4	165	36.1	0.002	0.012
5	48	10.7	0.000	0.031

for the training and test sets, respectively, where the error rate is the total sum of square errors divided by the total number of scenarios in the set.

For two of the test experiment graphs, 3 and 5, the algorithm was able to find a perfect edge assignment for the training data in all three of the trials. In the other three experiment graphs, the algorithm maxed out the iterations in some or all of the three trials. There does not appear to be a clear relationship between the combined total number of levels and rules under consideration for the algorithm and the test set error. In general, it appears that the total time it takes to solve is less when more rules are included. However, Experiment 4 highlights that this is not always the case.

3.4 Solving for Critical Nodes

The goal of creating a dependency representation is often to estimate how impacting (i.e., reducing the value of node level) various nodes in the graph impacts the overall mission. However, it can be used more broadly: Identifying node(s) that, if attacked, significantly impact the mission suggests where mitigations should be sought to harden these weak points or lessen the mission's dependency on them. Often, the leaf nodes in the mission representation represent physical components such as cyber assets. These might be the only points which an adversary may attack. Therefore we next introduce a simple greedy algorithm to identify the N most critical leaf nodes in a given network described in the general dependency representation framework, where N is an input to the algorithm. While this chapter assumes that the network is such that non-leaf nodes cannot be attacked, if this were not the case, the class of nodes to search for could be easily augmented.

In Sect. 3.4.1, a greedy algorithm is presented that identifies a set of critical nodes. In Sect. 3.4.2 the algorithm is applied to the illustrative example of Fig. 3.3 under two different assumptions. In Servi and Garvey (2017) and Chap. 1, the criticality of a node is implicitly defined as the derivative of the mission metric value with respect to the node metric value of a particular node, and both Servi and Garvey (2017) and Chap. 1 provide a fast algorithm to compute it. In principle, the associated fast algorithm in Servi and Garvey (2017) could be adapted to the assumed small number of discrete values, but the algorithm presented in this chapter more naturally deals with the discrete values.

3.4.1 A Greedy Algorithm for Identifying Sets of Critical Nodes

To solve the problem of finding a set of N nodes that impacts the mission more than any other set of N nodes, we have used a simple greedy algorithm. The algorithm first finds the leaf node that, when its node metric value is set to zero, causes the greatest negative impact on the mission. This is found by calculating the mission metric for all instances when any one leaf node metric value is set to zero (i.e., it is made inoperable) and choosing the leaf node(s) that minimizes the mission metric. For each subsequent iteration, the metric value of the critical nodes that the algorithm found previously is fixed at zero, and new leaf nodes are chosen by finding the minimum mission metric values for the cases when each of the remaining leaf nodes is also set to zero. The node that minimizes the mission metric at each iteration is added to the solution until it consists of N total nodes.

If there is more than one node that, when its metric value is set to zero, minimizes the mission metric, each of these options is added to a copy of the previous iteration's solution and considered separately as potential solutions. This can lead to a large set of candidates under consideration at each iteration. If no new leaf node reduces the mission metric value beyond that attained in the previous iteration, but fewer than N nodes have been found, then all options of new leaf nodes are considered as minimizing the mission metric value equally. Therefore, each of the remaining leaf nodes is added to the current solution separately to create a number of solutions to be considered in the next iteration. The algorithm stops when there are N nodes in the solution. However, if there are any previous solutions that have fewer nodes but the same impact on the mission metric value, the smallest set of nodes is output as the solution.

The pseudo-code for this algorithm is displayed in Algorithm 2 in the Appendix. The inputs are the network and the total number of critical nodes to be identified, and the output is the set of critical nodes and the resulting mission metric value if all of the nodes in this set are at the zero value. This algorithm deals with two sets: `activeAttackOptions` and `inactiveAttackOptions`. The `activeAttackOptions` consists of possible solution node sets to which new nodes can be added as the algorithm progresses. The `inactiveAttackOptions` consists of sets of nodes that do not reduce the mission metric value when another node metric value is set to zero. At the end of iteration i, every node set in `activeAttackOptions` of size $i-1$ is either stored in `inactiveAttackOptions` or discarded, so the next iteration's `activeAttackOptions` candidates are all of size i. Once stored in `inactiveAttackOptions`, a node set will not have new nodes added and will not be accessed again, until we are searching through all of the `activeAttackOptions` and `inactiveAttackOptions` candidates to output the smallest node set that has our determined minimum mission metric value.

The algorithm also references the function SINGLENODEATTACK(), which has inputs of a network and a set of "fixed nodes." The function returns the leaf node (or nodes in the case of a tie) that, if set to zero, will have the greatest negative impact on the mission metric value, given that all of the nodes in the "fixed nodes" set are fixed at the zero. The full pseudo-code for this function is in the Appendix.

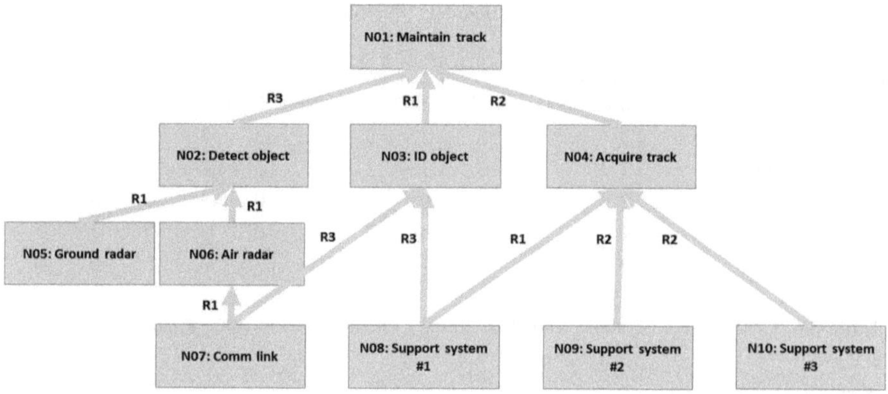

Fig. 3.4 An illustrative example of a dependency relationship with RiskMAP parameters

3.4.2 An Illustrative Example

Consider the dependency graph in Fig. 3.4. Each non-leaf node is labeled with a name and its m indicator, which is equal to 1 if the node aggregates with a maximum and -1 if the node aggregates with a minimum. Hence, as with RiskMAP (Watters et al. 2009), each node can take on one of the four discrete values: 0, 1, 2, and 3. These could be thought of as denoting failure, degradation, work-around, and operational levels, respectively. The (r_1, r_2) values are labeled at each edge. Each r_1 value is an index for a row in **B**, and each r_2 value is an index for an element in $\vec{a}$, defined as follows:

$$\vec{a} = \begin{bmatrix} 0 \; 1 \; 2 \; 3 \end{bmatrix} \qquad \mathbf{B} = \begin{bmatrix} 0 \; 1 \; 2 \; 3 \\ 1 \; 1 \; 3 \; 3 \\ 0 \; 0 \; 1 \; 3 \\ 2 \; 3 \; 3 \; 3 \end{bmatrix}.$$

While motivated in part by RiskMAP, it is not RiskMAP. The choice of **B** above is distinct from the **B** of RiskMAP and was chosen for illustrative purposes. Assuming that all leaf nodes are fully operational, i.e., ℓ_j is initialized at 3 for the four leaf nodes j, and that an attack on a leaf node causes it to enter a failure state, ℓ_j=0, the problem is to identify N nodes to attack that have the greatest negative impact on the level of the mission node, N01.

Table 3.5 shows the critical nodes determined from the greedy algorithm assuming that 0, 1, 2, or 3 nodes are attacked and the corresponding metric values for all nodes in the graph. The mission metric value, the top node N01, is in bold for each case. Note that for this network, the results of the algorithm happen to be optimal in terms of finding the node set that will minimize the mission node value given. However, this optimality is not guaranteed.

To illustrate the impact of the analysis on the dependencies rules, Table 3.6 reexamines the same network as Fig. 3.1, but this time using the choice of **B**

Table 3.5 Critical sets of nodes and their impact on the mission node if not functional

# of nodes attacked	Critical set	**Mission node, N01**	N02	N03	N04	N05	N06	N07	N08	N09	N10
0	N/A	**3**	3	3	3	3	3	3	3	3	3
1	N07	**2**	3	1	3	3	3	0	3	3	3
1	N08	**2**	3	0	3	3	0	3	0	3	3
2	N05,N08	**1**	1	0	3	0	0	3	0	3	3
3	N07,N09,N10	**0**	3	1	0	3	3	0	3	0	0

Table 3.6 Critical sets of nodes and their impact on the mission node assuming RiskMAP dependencies

# of nodes attacked	Critical set	**Mission node, N01**	N02	N03	N04	N05	N06	N07	N08	N09	N10
0	N/A	**0**	0	0	0	0	0	0	0	0	0
1	N07	**2**	0	1	3	0	0	3	0	0	0
1	N09	**2**	0	0	2	0	0	0	0	3	0
1	N10	**2**	0	0	2	0	0	0	0	0	3

corresponding to the RiskMAP (Watters et al. 2009) dependencies as specified in Fig. 3.4. Each edge is labeled with the rule, i.e., the row of **B**, characterizing the relationship between its operability value and the operability value of its parent node. As noted in Table 3.2, RiskMAP has no multi-child contribution aggregation rule, and the MaxMin rule, m_j, is always 1. So, without loss of specificity, neither is explicitly listed in Fig. 3.4. The results of the greedy algorithm are shown in Table 3.6. For RiskMAP dependencies, recall that the node metric values {0,1,2,3} correspond to the node being in the state of *operational*, *work-around*, *degradation*, and *failure*, respectively, so impairing a node corresponds to setting the value to 3. From Table 3.6, one sees that there are three different single-node attacks that degrade the network, yet there are no multiple node attacks that put the mission in a failure set.

What Tables 3.5 and 3.6 illustrate is that even with the same mission dependency map, the differences in the assumed details of the dependency relationship, through the **B** matrix, result in substantially different conclusions regarding the most critical nodes. In short, an accurate **B** sometimes matters.

3.4.3 Ties to Digital Twins

Digital twins (DTs) are a maturing capability increasingly adopted across domains such as advanced manufacturing, health care, automotive, smart cities, supply chains, and defense (Grieves 2002; Sharma et al. 2022) which aims to provide virtual representations of physical systems to support a system's life cycle in real time. Leveraging instrumented assets that stream telemetry data on the operational state and performance to the DT, DTs then apply predictive analysis (e.g., simulation, forecasting) to anticipate the physical system's future behavior. As DT use cases grow in scope and complexity, reliance on manual inputs from subject matter experts (SMEs) becomes inefficient and time-consuming. Emerging AI methods can assist DT construction and upkeep by extracting structure from documentation, learning parameters from telemetry, and automating model updates (Grieves and Hua 2024). The relationship between DTs and the methods developed in Sects. 3.3 and 3.4 can be summarized along four dimensions:

- Granularity: DTs support multi-scale representations from component level to often mirroring the physical system's detail. By design, this chapter's models operate at a coarser, mission-level granularity to support tractable performance analysis and criticality analyses.
- Inputs required for construction: DTs have substantial input for their construction. DTs need requirements and engineering documentation, live telemetry, SME knowledge, and, when live telemetry is not available, synthetic data. In contrast, Sect. 3.3 relies on simpler designed experiments that each vary component operability and report the resulting mission effectiveness.
- Outputs: DTs can generate high-fidelity states and telemetry including, for example, inputs suitable for Sect. 3.3. Section 3.3 produces dependency parameters that (i) enable high-level assessment of mission effectiveness and (ii) feed criticality analyses such as those in Sect. 3.4.
- Real-time dynamics: DTs monitor and assess streaming data and can detect structural system changes. Section 3.3, as currently designed, is a static analysis.

3.5 Conclusion

This chapter offered three new contributions:

(1) It provided a new general, flexible, and quantitative discrete framework for mission dependency representations, which allowed the user to specify a flexible number of discrete node operability values and dependency relationships, as was appropriate for a mission of interest. This new framework captured, as special cases, several previously proposed existing dependency representations.
(2) It provided an algorithm that took as input many scenarios of potentially failed nodes and the resulting mission value, perhaps from simulation experiments of a mission network, and inferred the mission dependency representation parameters most consistent with the experiments. This circumvented or complemented the more time-consuming use of a subject matter expect, an SME, to identify the dependency relationships.
(3) Finally, it provided an algorithm that took the mission dependency mapping and parameters as input and identified the critical nodes that could be used for defensive or offensive purposes.

Appendix A: The Genetic and Greedy Algorithms

Algorithm 1 Genetic algorithm for system identification

```
function GeneticAlgorithm(network, candidates, maxIterations, minThreshold, crossOverProbability, mutationProbability, scenarioInputs, scenarioMissionValues)
    i ← 0
    minError ← ∞
    bestCandidate ← {}
    while i < maxIterations do
        candidateError[candidate] ← 0 ∀candidates
        for each candidate do
            update network with candidate edge rules
            for each input scenario do
                calculatedMissionValue ← CalcMissionNode(network, scenarioInputs)
                scenarioMissionValue ← output mission value from scenario data
                candidateError[candidate] ← candidateError[candidate] + (calculatedMissionValue − scenarioMissionValue)^2
            end for
        end for
        minError ← min(candidateError)
        if minError < minThreshold then
            break while
        end if
        meanError ← mean(candidateError)
        selected ← {}
        for each candidate do
            if candidateError[candidate] < meanError then
                Add the candidate to selected int(meanError / candidateError[candidate]) times
            end if
        end for
        if the size of selected is less than the original number of candidates then
            randomly select more candidates to add to selected
        end if
        randomly draw, with replacement, ⌊|selected| / 2⌋ pairs of from the selected candidates
        for each pair do
            Crossover each edge rule between the pairs with probability crossOverProbability
        end for
        Results of crossovers are now called offspring
        for each offspring do
            Replace each edge rule in the offspring with a randomly selected edge rule with probability mutationProbability
        end for
        candidates ← offspring
        i ← i + 1
    end whilereturn candidates with minimum error
end function
```

Algorithm 2 Greedy algorithm for finding critical nodes

```
function GreedyCriticalNodes(network, N)
    activeAttackOptions ← {}
    inActiveAttackOptions ← {}
    [targetNodes, minMissionValue] = SingleNodeAttack(network, ∅)
    for each node n in targetNodes do
        add (n, minMissionValue) to activeAttackOptions
    end for
    for i in [2,3,...,N] do
        for (prevTargetNodes, prevMinMissionValue) in activeAttackOptions do
            [newTargetNodes, newMinMissionValue]= SingleNodeAttack(network, prevTargetNodes)
            if newMinMissionValue= prevMinMissionValue then
                Add (prevTargetNodes, prevMinMissionValue) to inActiveAttackOptions
            end if
            for node n in newTargetNodes do
                if prevTargetNodes ∪ node n not in activeAttackOptions then
                    Add (prevTargetNodes ∪ node n, newMinMissionValue) to activeAttackOptions
                end if
            end for
            Remove (prevTargetNodes, prevMinMissionValue) from activeAttackNodes
        end for
    end for
    greedyMissionValue ← minimum mission value in activeAttackOption
    greedyAttackNodes ← the minimum length node set(s) in activeAttackOption ∪ inActiveAttackOptions for which the mission value = greedyMissionValue
    return greedyAttackNodes, greedyMissionValue
end function
```

Appendix B: Functions for Algorithms

The CalcNodeLevels() function below is called in the SingleNodeAttack() function, which is used in Algorithm 2.

Algorithm 3 Algorithm to compute the node metric values

```
function CalcNodeLevels(network, targetNodes)
    nodeLevels ← {}
    orderedNodes ← list of nodes in network ordered such that the children of any node n are in the list before n
    for node n in orderedNodes do
        if node n is in targetNodes then
            Add (n, 0) to nodeLevels
        else if node n is a leaf node then
            Add (n, baseline level of node) to nodeLevels
        else
            aContribution ← 0
            aDenominator ← 0
            bContribution ← {}
            for each child c of node n do
                Add a_{r1(c,n)} ℓ_c to aContribution
                Add a_{r1(c,n)} to aDenominator
                Add b_{r2(c,n),ℓ_c} to bContribution list
            end for
            if aDenominator > 0 then
                aContribution ← ⌊aContribution / aDenominator⌋
            else
                aContribution ← {}
            end if
            Append aContribution to bContribution to form list contributions
            if node n is a minimizing node then
                Add (n, min(contributions)) to nodeLevels
            else if node n is a maximizing node then
                Add (n, max(contributions)) to nodeLevels
            end if
        end if
    end for
    missionNodeLevels ← the value of nodeLevels for the mission node return missionNodeLevels
end function
```

```
function SingleNodeAttack(network, fixed)
    Set all level of all nodes in C to 0
    targetNodes ← {}
    minMissionLevel ← ∞
    missionLevels ← {}
    for all leaf nodes \ fixed in network do
        level of mission node ← CalcNodeLevels(network, leaf node ∪ fixed)
        Add (leaf node, level of mission node) to missionLevels
        if Level of mission node ≤ minMissionLevel then
            minMissionLevel ← Level of mission node
        end if
    end for
    for (leaf node, level of mission node) in missionLevels do
        if level of mission node = minMissionLevel then
            Add leaf node to targetNodes
        end if
    end forreturn targetNodes, minMissionLevel
end function
```

References

Ahuja RK, Magnanti LT, Orlin JB (1988) Network flows: theory algorithms, and applications. Prentice Hall, Upper Saddle River

Albert R, Heong J, Barabasi AL (2000) Error and attack tolerance of complex networks. Nature 406(6794):378–382

Albert R, Albert I, Nakarado GL (2004) Structural vulnerability of the North American power grid. Phys Rev E 69(2):025103

Anderson-Cook, CM (2005) Practical genetic algorithms: 1099–1099

Asimov, I (1982) Catch that rabbit. In: The complete robot. Doubleday, New York, p 271

Bertsimas D, Sim M (2004) The price of robustness. Oper Res 52(1):35–53

Binith SS, Sathya S (2012) A survey of Bio inspired optimization algorithms. Int J Soft Comput Eng (IJSCE) 2(2):137–151

Birnbaum ZW (1969) On the importance of different components in a multicomponent system. In: Krishnaiah PR (ed) Multivariate Analysis II. Academic Press, New York, pp 581–592

Boussaid I, Lepagnot J, Siarry P (2013) A survey on optimization metaheuristic. Inf Sci 237(2013):82–117

Browning TR (2001) Applying the design structure matrix to system decomposition and integration problems: a review and new directions. IEEE Trans Eng Manag 48(3):292–306

Browning TR (2016) Design structure matrix extensions and innovations: a survey and new opportunities. IEEE Trans Eng Manag 63(1):27–52

Chen, X (2016) Critical nodes identification in complex systems. Complex Intell Syst 1(1–4):37–56

Costa LdF, Rodrigues FA, Travieso G, Villas Boas PR (2007) Characterization of complex networks: a survey of measurements. Adv Phys 56(1):167–242

Davis L (1991) Handbook of genetic algorithms. Van Nostrand Reinhold, New York

Dorigo M, Blum C (2005) Ant colony optimization theory: a survey. Theor Comput Sci 344(2–3):243–278

Garvey PR, Pinto CA (2009) Introduction to functional dependency network analysis. iN: Proceedings of the second international symposium on engineering systems. Massachusetts Institute of Technology, Cambridge. http://esd.mit.edu/symp09/day3.html

Garvey PR, Pinto CA (2012) Advanced risk analysis in engineering enterprise systems. Chapman & Hall/CRC, Boca Raton. ISBN 9781439826140

Garvey PR, Pinto CA, Santos JR (2014) Modelling and measuring the operability of interdependent systems and systems of systems: advances in methods and applications. Int J Syst Syst Eng 5(1):1–24

Gevers M (2005) Identification for control: from the early achievements to the revival of experiment design. Eur J Control 11(4–5):335–352

Glover F (1986) Future paths for integer programming and links to artificial intelligence. Comput Oper Res 13(5):533–549

Goldberg DE, Deb K (1991) A comparative analysis of selection schemes used in genetic algorithms. In: Foundations of genetic algorithms. Morgan Kaufmann, California, pp 69–93

Grieves M (2002) Completing the cycle: using PLM information in the sales and service functions [slides]. SME Management Forum (19). Troy, MI, USA, Oct 31

Grieves M, Hua EY (2024) Digital Twins, simulation, and the metaverse: : driving efficiency and effectiveness in the physical world through simulation in the virtual worlds. Springer Nature, Berlin

Grindal M, Offutt J, Andler SF (2005) Combination testing strategies: a survey. Softw Test Verif Reliab 15(3):167–199

Guariniello C, DeLaurentis D (2017) Supporting design via the system operational dependency methodology. Res Eng Des 28(1):53–69

Hastings G, Montella L, Watters J (2009) MITRE Crown Jewel Analysis Process. MITRE Technical Report MTR090088

Kuhn DR, Wallace D, Gallo AM (2004) Software fault interactions and implications for software testing. IEEE Trans Softw Eng 30(6):418–421

Kuhn DR, Lei Y, Kacker RN (2008) Practical combinatorial testing: beyond pairwise. IT Prof 10(3):19–23

Kuo W, Zhu X (2012) Importance measures in reliability, risk, and optimization: principles and applications. Wiley, New York

Ljung L (1998) System identification. In: Signal analysis and prediction. Birkhäuser, Boston, pp 163–173

Mitchell M (1996) An introduction to genetic algorithms. MIT Press, New York

Pinto CA, Garvey PR (2012) Advanced risk analysis in engineering enterprise systems. CRC Press, New York, pp 239–247

Reliability Information Analysis Center (2005) System reliability toolkit. Topic 6.6.2: Fault Tree Analysis (FTA). RIAC/Quanterion Solutions, New York

Schrijver A (1998) Theory of linear and integer programming. Wiley, New York

Servi LD, Garvey PR (2017) Deriving global criticality conditions from local dependencies using functional dependency network analysis (FDNA). Syst Eng 20(4):297–306

Sharma A., Kosasih E, Zhang J, Brintrup A, Calinescu A (2022) Digital Twins: state of the art theory and practice, challenges, and open research questions. J Ind Inf Integr 30:100383

Shen Y, Nguyen NP, Xuan Y, Thai MT (2013) On the discovery of critical links and nodes for assessing network vulnerability. IEEE/ACM Trans Networking 21(3):963–973

Ventresca M, Aleman D (2015) Efficiently identifying critical nodes in large complex networks. Comput Soc Netw 2(1):6

Walker A, Hallam J, Willshaw D (1993) Bee-havior in a mobile robot: the construction of a self-organized cognitive map and its use in robot navigation within a complex, natural environment. In: Proceedings of the ICNN'93–IEEE international conference on neural networks, vol. III. IEEE Service Center, Piscataway, pp 1451–1456

Wasserman S, Faust K (1994) Social network analysis: methods and applications. Cambridge University, Cambridge

Watters J, Morrissey S, Bodeau D, Powers S (2009) The Risk-to-Mission Assessment Process (RiskMAP): a sensitivity analysis and an extension to treat confidentiality issues. In: The MITRE Corporation/Institute for Information Infrastructure Protection (I3P), Pub Rel 09-2994, pp 25–26

Yu T-L, Goldberg DE (2004) Dependency structure matrix analysis: off-line utility of the dependency structure matrix genetic algorithm. In: Genetic and evolutionary computation conference (GECCO 2004), Part II. Lecture Notes in Computer Science, vol 3102. Springer, Berlin, pp 355–366

Part II
Mission Dependency Network Analysis with Physical and Functional Components

Chapter 4
Disrupting Functional Pathways Having Supporting Physical Systems: An OR Approach

We have two eyes to see two sides of things, but there must be a third eye which will see everything at the same time ....

D. T. Suzuki Humphreys (1985)

4.1 Introduction

Chapter 1 introduced an approach to modeling a mission functionality dependent network consisting of only functional components, i.e., components that are activities in support of a mission. Chapters 2 and 3 enlarged the mathematical possibilities for how such mission networks can be modeled and analyzed: Chap. 2 did this with a continuous variable framework, and Chap. 3 did this with a discrete variable framework. This chapter takes a "third eye" view and enlarges the Chap. 1 analysis in a different direction using Operations Research (OR) methods: It acknowledges that in some mission planning and analysis, a mission success depends on the ability to accomplish multiple activities, which in turn depend on the functionality of supporting physical systems. These physical systems may lose functionality due to, for example, cyberattack, natural disaster, or aging infrastructure. It is therefore imperative to understand and characterize the risk that each system poses to overall mission effectiveness. Furthermore, this chapter acknowledges and addresses the realist possibility that the parameters that characterize the mission are not known perfectly.

There are several frameworks to analyze nodal criticality, depending on the intended use case. Graph analysis can be applied to identify critical nodes based solely on network topology. Examples of this approach include determining cutsets (Ahuja et al. 1993; Stoer and Wagner 1997), degree of node centrality (Wasserman and Faust 1994), shortest paths (Albert et al. 2004), or pairwise connectivity (Shen et al. 2013). Similar methods arise in reliability theory, whereby the network functionality is determined as a logical function of the node operability. Criticality measures include Birnbaum Structural Importance (Birnbaum 1968), Fussell-Vesely importance (Fussell 1975; Vesely 1970), and Risk Reduction Worth and Risk

L. Servi, *Mission Dependency Network Analysis*, Risk, Systems and Decisions,
https://doi.org/10.1007/978-3-032-11260-6_4

Achievement Worth (van der Borst and Schoonakker 2001). Alternative approaches arise when supplementing the network topology with functional dependency relationships between nodes, e.g., Functional Dependency Network Analysis (FDNA) (Garvey et al. 2009; Servi and Garvey 2017) and System Operational Dependency Analysis (SODA) (Guariniello and DeLaurentis 2017).

In this chapter, we investigate an alternative approach to identifying critical nodes specific to networks that contain *both* physical and functional components. We use the term *hybrid mission network* to refer to networks where each node is designated as either *physical* or *functional*. Each node will be in a binary state of operability or inoperability (as opposed to, e.g., FDNA or SODA where continuous operability scores can be assigned). The distinction between a physical node and a functional node determines how the node can fail (i.e., it becomes inoperable). Physical nodes can fail independently or be disabled. The operability of a functional node will depend on the operabilities of certain subsets of physical nodes, called *supporting combinations*, and other functional nodes, if any, it depends on.

The hybrid mission network can naturally model missions where the success of the functions to be carried out depends on the operabilities of physical assets. Although hybrid mission networks can be applied more broadly (e.g., to computer systems or networks of infrastructure), this chapter will focus on mission networks as the primary application. A mission network is a hybrid mission network, with one or more nodes designated as the start node and a single node designated as the finish node. We say that the mission is successful if there is a path of functional nodes, all of which are operable, from the mission start to the finish. Figure 4.1 shows a simple use case constructed for illustrative purposes, where the mission network models a retailer's process to check out a customer.

In this example, there are two paths of functional nodes that lead to mission success, depending on if the customer is paying with cash or with a credit card. Each physical node represents the hardware or software that enables a given functional node. For instance, the functionality of "Determine item prices" requires the physical node "Barcode scanner." In the case where multiple physical nodes are

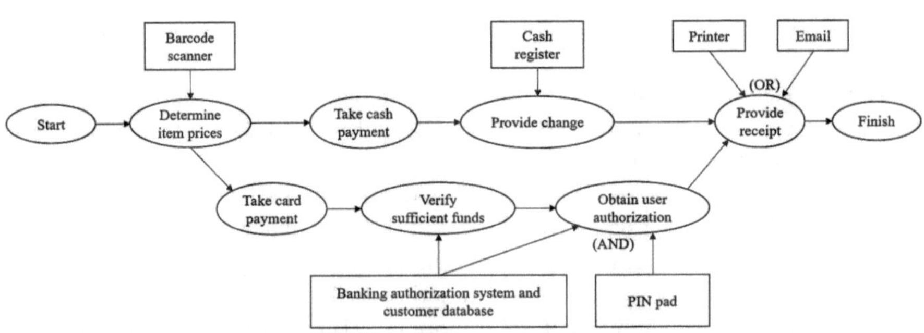

Fig. 4.1 A hybrid mission network modeling the functions that a retailer must perform to check out a customer. Functional nodes are represented as ovals, and the physical nodes are represented as rectangles

related to a single functional node, the relationships will include a set of logical operations. For example, a receipt can be provided physically with a printer or electronically via email. Therefore the functional node "Provide receipt" is operable if physical node "Printer" or "Email" is operable. In contrast the functional node "Obtain user authorization" requires both the PIN pad and the banking authorization system to be operable. Although this example only contains logical AND and OR functions, the proposed framework allows us to consider more complex logical functions relating the operability of each functional node to the set of physical nodes. The network can contain functional nodes that do not rely on physical nodes, for example, the nodes associated with taking cash or card payment. In practice, these nodes will not impact the analysis but can be included if their presence lends additional clarity to the process or function under consideration.

For these hybrid mission networks, this chapter uses an optimization-based approach to determine the physical nodes that are most critical. This optimization is posed from the perspective of an attacker who aims to disable some subset of physical nodes to prevent mission success. Adversarial models have been used to identify critical components in key infrastructure such as oil pipelines, electrical grids (Brown et al. 2006), and road networks (Bell et al. 2008). These are typically evaluated through bi-level optimization models (also known as "attacker-defender" models). For the context of this chapter, we are solely concerned with determining nodal criticality and not in the selection of explicit protective measures. This can be done with a simplified attacker-defender problem where the defender[1] has no recourse. Intuitively, the model then reduces to a single-level optimization problem for the attacker,[2] and the defender can only evaluate the attacker's potential attacks to understand nodal criticality. This understanding implicitly suggests mitigation options.

As some of the required data may not be known with complete certainty, we also propose a robust extension to the hybrid mission network that can account for uncertainty. This robust model must be structurally different from typical approaches, because the solutions to the optimization problem will be analyzed and interpreted through the viewpoint of the defender. Established methods for robust optimization such as Ben-Tal and Nemirovski (1999) and Bertsimas and Sim (2004) generate solutions that are robust against infeasibility. An attacker may be concerned with feasibility, but a defender should also consider attacker solutions that have a low probability of being feasible but have high impact. The robust optimization framework in Sect. 4.2.2 will allow a defender to evaluate these solutions for varying levels of feasibility probabilities. This reflects varying levels of conservatism of the defender.

[1] As noted in Brown et al. (2006), the term *defender* can be a misnomer in general and let alone when no defensive actions are available. However, it serves to describe an entity whose objective is diametrically opposed to that of the attacker. In this context, the defender is the party who benefits from mission success and physical node operability.

[2] The terms *adversary* and *attacker* will be used interchangeably in this chapter.

The remainder of this chapter is organized as follows. Section 4.2 presents the notation used throughout the chapter and formally defines the problem being solved. The optimization problem is then solved. Section 4.3 contains examples to illustrate first in the context of perfect information, and then in a robust framework, accounting for areas of possible uncertainty and a brief discussion regarding how this analysis can be reinterpreted in terms of a defender making decisions which anticipate an attacker's later actions. Additionally, this section includes a brief discussion of solution performance. Lastly, Sect. 4.4 summarizes the chapter.

4.2 Problem Definition

4.2.1 Notation and Definitions

A mission network is a directed acyclic graph (DAG) composed of functional nodes, including a single "Finish" node. These nodes may represent activities that need to be accomplished in a certain order, dictated by the network arcs, for the mission to be deemed successful. In this chapter the mission network will be modeled by a hybrid mission network composed of physical nodes and functional nodes. The physical nodes represent systems or tools to support a functional node to perform its task. The functional nodes may or may not be supported by physical nodes. Disabling a physical node would propagate up to impact the performance of the supported functional node, potentially preventing mission completion. More generally, a functional node may rely on sets of physical nodes and arbitrary logical relationships among them. Regardless of its type, a node is in a binary state of operable or inoperable. There always exist a single "Finish" functional node and one or more "Start" functional nodes in such a network model.

We use the following notation to capture the sets and relationships that characterize the mission:

- $\mathcal{I}$ is the set of functional nodes.
- $\mathcal{J}$ is the set of physical nodes.
- $\mathcal{P}(i)$ is the set of functional nodes which directly precede functional node i in the mission. These sets define the topology of the functional node network. This partitions $\mathcal{I}$ into two subsets:
  $\mathcal{I}_0 = \{i \in \mathcal{I} : \mathcal{P}(i) = \emptyset\}$. This can be interpreted as the set of "Start" nodes.
  $\mathcal{I}_+ = \{i \in \mathcal{I} : |\mathcal{P}(i)| > 0\} = \mathcal{I} - \mathcal{I}_0$
- $C_k(i) \subset \mathcal{J}$ is the kth set of physical nodes. The functional node i is operable if all physical nodes in $C_k(i)$ are operable for some value of $k \in \{1, 2, \ldots, n_i\}$. We refer to $C_k(i)$ as the *supporting combinations* of functional node i.
- n_i is the number of distinct supporting combinations of physical nodes for the functional node i.

- If the operability of functional node i does not depend on the operability of any physical nodes, we use the convention that $n_i = 1$ with $C_1(i) = \emptyset$.
- c_j is the cost to disable physical node j.
- t_j is an impact coefficient for physical node j. This can capture an attacker's benefit from attacking j and separate from the impact that it has on the mission network. We assume that $t_j \geq 0$.

We frame an optimization problem to determine which physical nodes are the most critical. This problem is posed from the viewpoint of an attacker with perfect information. The attacker can choose the physical nodes to disable based on the individual impacts of removing nodes, as well as secondary network effects. This requires defining the additional quantities:

- $x_j \in \{0, 1\}$ is a binary decision variable denoting whether or not to render inoperable the physical node $j \in \mathcal{J}$. If $x_j = 1$, we say that physical node j is "disabled" or "attacked." If $x_j = 0$, then physical node j is still operable.
- $r_i \in \{0, 1\}$ is a binary variable for each functional node $i \in \mathcal{I}$. This will equal one if the following two conditions are true:

  1. There exists a path of functional nodes, all operable, from some start node $s \in \mathcal{I}_0$ to node i.
  2. At least one supporting combination of physical nodes $C_k(i)$ is operable for some k.

  When both these conditions are met, the functional node i is operable, and it is reachable from $\mathcal{I}_0$ through a path of operable nodes. Otherwise, r_i will equal zero. We use an index of zero to refer to the final functional node in the network (sometimes also called the "Mission Node"). The mission is considered successful if $r_0 = 1$.

- B is a nonnegative value representing the attacker's resources, hereafter referred to as a budget.
- α is a nonnegative constant corresponding to the value of mission success, relative to the benefits of disabling individual systems. As α increases, disabling the mission becomes more important than any byproduct benefit from disabling a specific physical node.

For a fixed budget B, the optimal solution to the following problem would determine the set of physical nodes that an attacker should disable to maximize their overall benefits, t_j and potentially disrupt the mission network, r_0. Any physical nodes that are disabled can be considered critical for this level of attacker budget.

$$\min \left[\alpha r_0 - \sum_{j \in \mathcal{J}} t_j x_j \right] \tag{4.1}$$

subject to

$$\frac{1}{|\mathcal{P}(i)|}\sum_{m\in\mathcal{P}(i)} r_m - \sum_{j\in C_k(i)} x_j \le r_i, \qquad \forall i \in \mathcal{I}_+,\ k = 1,\cdots,n_i, \tag{4.2}$$

$$1 - \sum_{j\in C_k(i)} x_j \le r_i, \qquad \forall i \in \mathcal{I}_0,\ k = 1,\cdots,n_i\,, \tag{4.3}$$

$$\sum_{j\in\mathcal{J}} c_j x_j \le B\,, \tag{4.4}$$

$$x_j,\ r_i \in \{0,1\}, \qquad \forall\, j \in \mathcal{J},\ \forall\, i \in \mathcal{I}. \tag{4.5}$$

The objective function (4.1) combines the two goals for the attacker. First, there is the aim to prevent mission success, and the relative value of doing so, captured in the first term αr_0. Second, the attacker is incentivized to disable as many physical nodes as possible, to gain the additional benefits from each t_j.

Constraint (4.2) enforces that a functional node i is operable under two conditions. First, there must be a functional node preceding it that is also operable, which would force the first summation over $P(i)$ to be positive but not greater than 1. Second, there must be at least one supporting combination $C_k(i)$ which is functional (i.e., none of the physical nodes within $C_k(i)$ were attacked). If one value of k satisfies this, then the second summation over $C_k(i)$ will be zero. When both conditions are satisfied, r_i will be forced to a value of 1 as $r_i \in \{0, 1\}$. The same logic also applies to constraint (4.3), except the nodes in $\mathcal{I}_0$ do not have any precedents.

Constraint (4.4) ensures that the set of physical nodes attacked does not exceed the attacker budget. Note that the output of this problem is budget-specific, so increasing the value of B and solving it will show how the set of most critical nodes may change when dealing with attackers that are increasingly well resourced. This process can be performed iteratively, starting with $B = \min_j c_j$ and slowly increasing to $B = \sum_j c_j$.

The output of this analysis will be a sequence of (B, x) pairs that are Pareto-efficient with respect to budget and objective value. This sequence characterizes the attacker's optimal strategy for all levels of budget. It contains the number of times each physical node is attacked across multiple budgets, as well as the lowest level of budget before the attacker should disable that node. Both of these quantities would give insight into physical node criticality.

This formulation can be viewed as a combination of a knapsack problem and a layered network interdiction problem. If $\alpha = 0$, constraints (4.2) and (4.3) can effectively be ignored as the values of the r_i's do not impact the objective, so the formulation becomes a standard 0-1 knapsack problem. If $t_j = 0$ for all j, then physical nodes will only be attacked to disable a set of functional nodes in order to separate the mission into unconnected components. This separation of the network is equivalent to interdicting flows with unit capacities on the arcs between functional nodes (Wood 1993).

4.2.2 Robust Formulation

The optimization problem in Sect. 4.2.1 permits the defender to anticipate the attacker's actions based on perfect information of the physical node attributes. However, in practice, this is not a realistic assumption, as the required data is not necessarily known with absolute certainty. A more desirable framework would enable the defender to hedge against what is known to be unknown and to generate solutions with differing levels of conservatism. There are several robust optimization methods that can incorporate considerations for unknown data, depending on the structure of the uncertainty and desired level of solution conservatism (Ben-Tal and Nemirovski 1999; Bertsimas and Sim 2004; Soyster 1973).

In this section, we propose a robust counterpart to the original optimization problem to meet these ends. As will be discussed, this formulation is also equivalent to solving a different problem, where the precise parameters are known but the defender can employ countermeasures to increase the cost or decrease the benefit of a node being attacked.

Consider first the situation where the cost to disable a physical node j is between a known upper bound and lower bound, with the exact cost value unknown:

$$c_j \in \left[c_j^\ell, c_j^\ell + \Delta_j^c\right], \qquad \Delta_j^c \geq 0.$$

The most conservative strategy for a defender is to assume that every physical node can be disabled with its lower bound cost, c_j^ℓ. This scenario assumes the attacker will require the least amount of resources possible to disable any physical node. However, this strategy may lead to solutions that are overly pessimistic and therefore unrealistic. As an alternative, we introduce a parameter Γ_c that will serve as an upper bound to the number of physical nodes that can be disabled with their worst-case cost value c_j^ℓ. The remaining $|\mathcal{J}| - \Gamma_c$ physical nodes will instead require their upper bound cost values of $c_j^\ell + \Delta_j^c$ to be disabled. Setting $\Gamma_c = |\mathcal{J}|$ is equivalent to the most conservative scenario.

To implement this more robust model, we introduce variables y_j that will indicate if a physical node j can be disabled for its lowest cost. That is, when $y_j = 1$, physical node j can be disabled with the lower bound cost c_j^ℓ. Alternatively, $y_j = 0$ corresponds to the attacker needing to spend the upper bound cost of $c_j^\ell + \Delta_j^c$ to disable node j. Although we assume $0 \leq y_i \leq 1$, it can be shown that there will always exist an optimal solution where $y_i \in \{0, 1\}$, if Γ_c is an integer. Hence, the robust model entails replacing (4.4) with constraints (4.6), (4.7), and (4.8):

$$\sum_{j \in \mathcal{J}} \left((c_j^\ell + \Delta_j^c) x_j - \Delta_j^c y_j \right) \leq B, \tag{4.6}$$

$$0 \leq y_j \leq x_j, \qquad \forall j \in \mathcal{J}, \tag{4.7}$$

$$\sum_{j \in \mathcal{J}} y_j \leq \Gamma_c. \tag{4.8}$$

The constraints above will force up to Γ_c physical nodes to have their effective cost to disable at the lower bound. Since the y_j's are decision variables in the optimization problem, the subset of Γ_c nodes will be chosen to maximize the benefit of the attacker. Constraint (4.7) ensures that the lowest cost can only occur for nodes that are attacked, which prevents the model from automatically setting $y_j = 1$ for nodes j with the highest cost spread Δ_j^c. This is a simple and intuitive modification to the original optimization problem, which takes advantage of the fact that x_j is binary.

This approach is based loosely on the methodology of Bertsimas and Sim (2004), although there are a few key differences. While both approaches assume an interval of uncertainty and bound the number of extreme values with a user-defined parameter, the methodology in Bertsimas and Sim (2004) is based on generating solutions that are robust against infeasibility (i.e., solutions that are likely to remain feasible if cost assumptions differ from the estimated values c). On the other hand, constraints (4.6)–(4.8) are a way to hedge against the possibility that the nominal solution is suboptimal (i.e., the attacker can actually inflict more damage than expected by using assumed cost values c). The former approach defines a robust solution from the attacker's perspective and the latter from the perspective of the defender red teaming the attacker's actions. A conservative defender would still want to analyze those attack strategies that are likely to be infeasible but have a high impact on the objective function.

An analogous approach can be used to incorporate robustness around the physical node impact scores. We assume that uncertain impact scores lie in the known range:

$$t_j \in \left[t_j^u - \Delta_j^t, t_j^u\right], \qquad \Delta_j^t \geq 0. \tag{4.9}$$

From the viewpoint of an attacker, the best-case scenario occurs when t_j equals its upper bound t_j^u. In this scenario, a physical node failure would create the greatest benefit. In a similar manner as above, we introduce variables v_j to indicate when node j has the lower bound impact score and limit the number of these occurrences to Γ_t. Only the remaining $|\mathcal{J}| - \Gamma_t$ physical nodes will have impact scores equal to the upper bound t_u^ℓ.

The following model restates the problem formulation, factoring in robustness around both physical node cost and impact. Given the same data as listed in Sect. 4.2.1, as well as the known ranges of uncertainty around the values of t_j and c_j, we aim to solve the following problem:

$$\min\left[\alpha r_0 - \sum_{j\in\mathcal{J}}\left(t_j^u x_j - \Delta_j^t v_j\right)\right] \tag{4.10}$$

subject to

$$\frac{1}{|\mathcal{P}(i)|}\sum_{m\in\mathcal{P}(i)} r_m - \sum_{j\in C_k(i)} x_j \le r_i, \qquad \forall i \in \mathcal{I}_+,\ k = 1, \cdots, n_i, \tag{4.11}$$

$$1 - \sum_{j\in C_k(i)} x_j \le r_i, \qquad \forall i \in \mathcal{I}_0,\ k = 1, \ldots, n_i, \tag{4.12}$$

$$\sum_{j\in\mathcal{J}} (\ (c_j^\ell + \Delta_j^c)\, x_j - \Delta_j^c\, y_j\ ) \le B, \tag{4.13}$$

$$0 \le y_j \le x_j, \qquad \forall j \in \mathcal{J}, \tag{4.14}$$

$$0 \le v_j \le x_j, \qquad \forall j \in \mathcal{J}, \tag{4.15}$$

$$\sum_{j\in\mathcal{J}} y_j \le \Gamma_c, \tag{4.16}$$

$$\sum_{j\in\mathcal{J}} v_j \le \Gamma_t, \tag{4.17}$$

$$x_j,\ r_i \in \{0, 1\}, \qquad \forall j \in \mathcal{J}, \forall i \in \mathcal{I}. \tag{4.18}$$

Observe that we could also treat both Γ_c and Γ_t as variables without changing the linear structure of this problem. Therefore, if desired, one can introduce the parameter Γ_{tot} to control the level of the total uncertainty by bounding the summation of the variables Γ_c and Γ_t:

$$\Gamma_t + \Gamma_c \le \Gamma_{\text{tot}}.$$

As noted earlier in this section, these equations could be viewed with a radically different but mathematically equivalent interpretation: Suppose there is no uncertainty, the cost to attack node j is c_j^ℓ, and the impact of doing so has a benefit of t_j^u. Suppose further that the defender wanted to and had the resources to do some mitigating actions, which would raise Γ_c of the costs of attacking node j from c_j^ℓ to $c_j^\ell + \Delta_j^c$ and lower Γ_t of the benefits of attacking node i from t_j^u to $t_j^u - \Delta_j^t$. Finally, suppose the defender acts first with mitigation actions in a manner that is visible to the attacker and before the attacker's decisions are made. The optimal actions of the defender and attacker can be determined by the same equations above.

4.3 Solution

4.3.1 Example and Discussion of Solution Representation

For purposes of simplicity and exposition, we illustrate our methods using a small example, although the methods apply to much larger networks. This will be shown

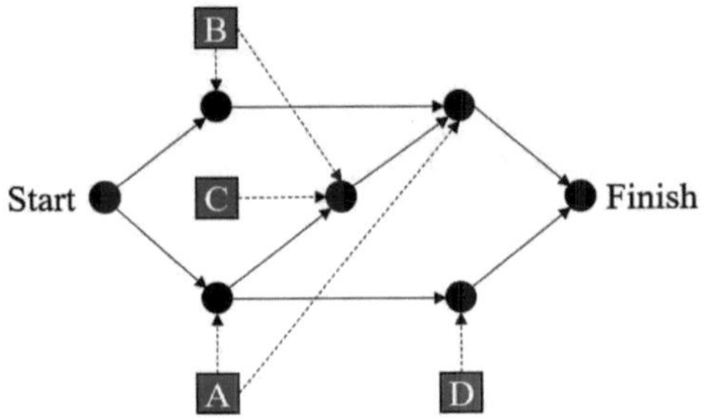

Physical Node (j)	Cost to Attack (c_j)	Node Impact (t_j)
A	2	1
B	2	2
C	1	3
D	1	4

Fig. 4.2 A small example of a hybrid mission network consisting of physical and functional nodes. The left image depicts the hybrid mission network, with functional nodes as circles, and physical nodes as squares. The table on the right captures the physical node attributes

in Sect. 4.3.2 on networks with thousands of nodes. An illustrative example of a hybrid mission network is shown in Fig. 4.2. The left portion of this figure depicts the network topology, where the functional nodes are circular and the supporting physical nodes are squares. In this example, when multiple physical nodes support the same functional node, we assume a logical AND relationship. The right portion of this figure shows the characteristics of each physical node, in terms of cost to attack and the impact to the objective of each node failing individually. In this example, we choose a network value parameter $\alpha = 20$, indicating that mission success (i.e., start-finish connectivity) is twice as important as the relative benefit of all systems failing individually ($\sum t_j = 10$).

To determine the physical nodes that are most critical, we solve the optimization problem described in Sect. 4.2.1 for a sequence of increasing budgets. This will generate a set of Pareto-efficient solutions that can in turn shed light on nodal criticality. Given the small size of this particular example, we can determine the optimal solutions for all values of budget. The left part of Fig. 4.3 is a table that shows the solutions to this problem when the budget takes integer values. Here, a black cell indicates which physical nodes were attacked in the optimal solution for the given budget. To the right of the table is a graph that shows the objective value for each of these solutions. For attacker budgets greater than or equal to 2, the mission network can be separated by targeting sets of physical nodes that support entire network cutsets causing r_0 to become 0. Since the objective function equals a weighting of r_0 minus any separate attacker benefit from attacking individual nodes, the right part of Fig. 4.3 shows the optimal objective value becoming negative as the budget exceeds 2.

The solutions captured in Fig. 4.3 are dependent on the attacker's budget. This is a useful representation if the physical node attack costs are known and the attacker's resources can be accurately estimated. However, in practice, the budget might be a numerical proxy for capability, and details of an attacker's capability might be unknown. In such cases, a more useful analysis output would be an ordering of physical nodes based on criticality that is independent of an adversary's budget. Unfortunately, there is not a single unambiguous way to perform such an ordering.

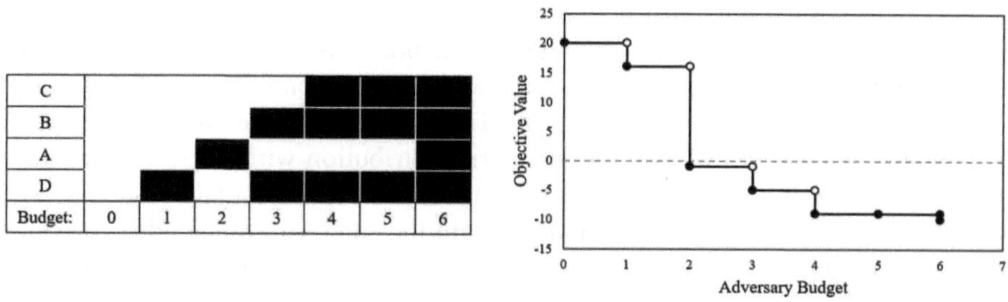

Fig. 4.3 The solution to the example in Fig. 4.2. The left table depicts the optimal solution for each budget by identifying in black the best choice of systems for an attacker to attack. For example, with a budget of 3, an adversary should attack physical nodes B and D. The graph on the right shows the corresponding Pareto frontier

We propose two basic heuristic methods to rank the physical nodes based on the set of Pareto-efficient solutions collected. The first approach is to rank physical nodes according to the solution in which they were attacked with the minimum budget. This is the approach taken when ordering the rows in the solution table of Fig. 4.3, where D is the most critical (as it was first in a solution when the budget was 1), followed by A, B, and C (as they were first in a solution when the budget was 2, 3, and 4, respectively). This ranking places an emphasis on physical nodes that can be disabled for the lower values of budget.

An alternate approach is to rank the physical nodes according to the number of times they appear in the set of optimal solutions. This ranking strategy captures the average solution behavior over a set of attackers with increasing values of budget. Using this logic, D remains the most critical as it was attacked five times, followed by B, C, and then A.

One might note that if there is a monotonicity in the figure on the left (i.e., if a node is the best to attack for a budget of B, it will also be the best to attack for all budgets more than B), these two rankings of the physical node criticality will be identical. The two rankings presented are not exhaustive; rather, they represent two simple approaches to solving this problem. There are other ways to define a ranking, given a solution diagram like that in Fig. 4.3.

4.3.2 Numerical Results

The methodology outlined above requires solving many optimization problems to generate a comprehensive set of solutions. In practice, the networks of interest may be orders of magnitude larger than the small network example presented in Sect. 4.3.1. To determine the numerical efficacy of our approach, we found the optimal solution for 75 generated test cases of different sizes, of different proportions of physical to functional nodes, and with different budget values. To limit our analysis to graphs of most practical interest, our test networks contain a basic underlying structure different from completely random DAGs.

We use the following construction to create a hybrid mission network test case, where the number of physical and functional nodes is specified as inputs: The hybrid mission network generation begins with a collection of sets of L functional nodes denoted by K_ℓ, for $\ell = 1, \ldots, L$. The number of nodes in each set, $|K_\ell|$, is randomly generated following a geometric distribution with a specified average μ_f. The network is initialized with all forward edges between adjacent sets, K_ℓ and $K_{\ell+1}$, and no edges between any nodes within the same set. Some edges are then removed with a given probability, p_{trim}, unless removing the edge would cause the in-degree or out-degree of any functional node to drop below one. By avoiding these cases we ensure that the network remains connected.

Finally, a randomly selected set of functional nodes is assigned to each physical node. The size of this set also follows a geometric distribution with average μ_p, implying that each physical node will support at least one functional node. If multiple physical nodes supported the same functional node, a logical "AND" relationship is assumed. One example of such a test case is shown in Fig. 4.4 before adding a "start" and "finish" node. The cost to disable each physical node and the impact data are generated from discrete uniform distributions. For the sake of simplicity, these test cases assume perfect information and therefore rely on solving the formulation from Sect. 4.2.1.

Following this method, a set of 75 test instances were generated, by increasing the size of the network, and varying ρ, the ratio of physical nodes to functional nodes. We solved instances with $\rho = |\mathcal{J}|/|\mathcal{I}| \in \{\frac{1}{4}, 1, 4\}$. Each test case was solved 100 times, where the right-hand side of (4.4) was increased from 1% of the total budget required to disable all physical nodes, up to 100%. The solutions to these 100 optimizations define a diagram like that pictured in Fig. 4.5 and should

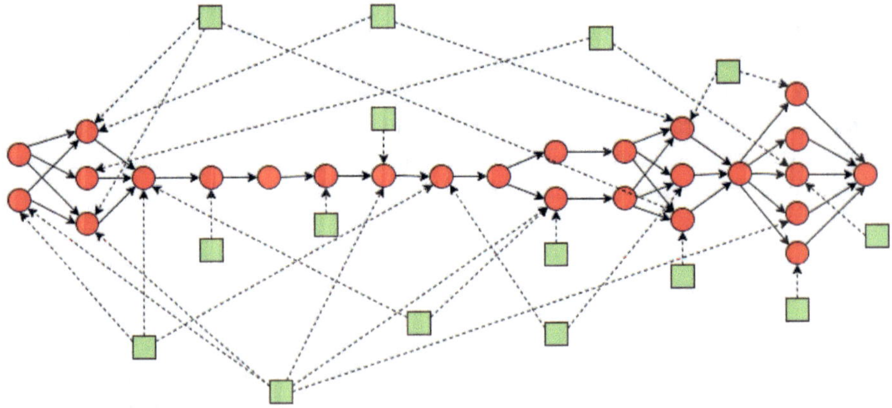

Fig. 4.4 An example of a generated test case, with 26 functional nodes (represented as circles) and 15 physical nodes (represented as squares). Other parameters were $\mu_f = 2.5$, $\mu_p = 2.5$, and $p_{\text{trim}} = 0.3$. Before using these randomly generated networks, a "start" node should precede the functional nodes that are currently not preceded by any other functional nodes, and a "finish" node should follow all functional nodes generated that are not followed by any functional nodes

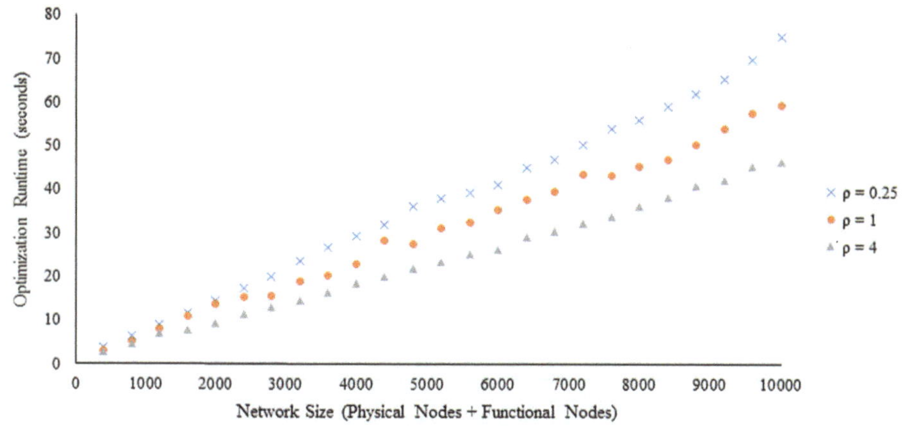

Fig. 4.5 A scatterplot depicting the relationship between optimization runtime and network size. Each point represents the total time to solve 100 optimizations, corresponding to increasing values of attacker budgets. The different series in the plot represent different distributions of physical and functional nodes

constitute a representative set of solution data to inform the discussion on physical node criticality. All test instances were implemented in Python 2.7 and solved with Gurobi 7.0 on a virtual machine with 64 GB RAM and 8 2.20 GHz Intel processors. The runtime for these experiments is shown in Fig. 4.5.

For each value of ρ, Fig. 4.5 shows an approximately linear relationship between the total optimization runtime and the size of the network. The slope of this line is dependent on the value of ρ, which characterizes the proportion of nodes which are physical versus functional. The test cases with lower values of ρ (corresponding to more functional nodes than physical nodes) typically took more time to solve. One possible explanation for this behavior is that the number of constraints scales with the number of functional nodes, so increasing the number of functional nodes may lead to a more challenging optimization problem.

This figure also shows that the approach considered is numerically tractable, as it takes roughly a minute to complete 100 optimizations on a network with 10,000 nodes. The time to solve an optimization is also nearly independent of the right-hand side of (4.4). Figure 4.6 shows the optimization runtimes (averaged over the set of 25 network instances of differing sizes), for each value of ρ and each right-hand side considered. Apart from the final optimization, where the budget is sufficiently high to disable every physical node, the solution times are roughly constant.

4.4 Conclusion

This chapter examined hybrid mission networks consisting of functional nodes representing, for example, activities needed for mission success, and physical nodes

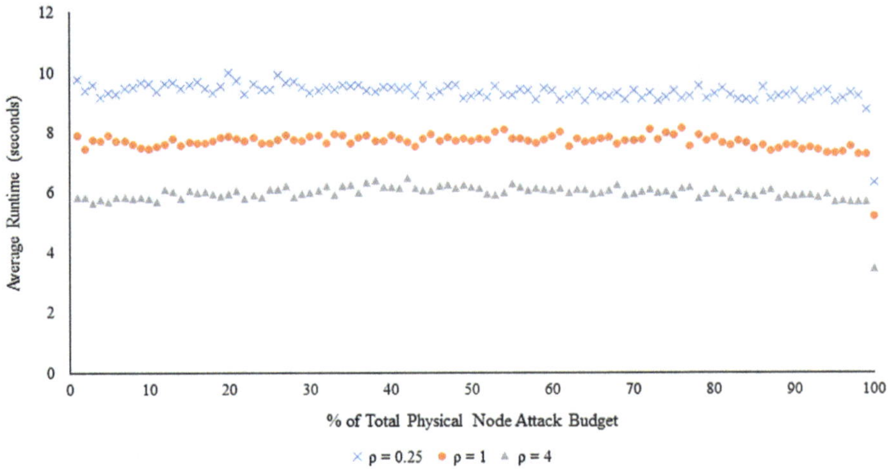

Fig. 4.6 A scatterplot showing the average optimization runtime relative to the increasing budget constraint, for different values of ρ

representing, for example, systems or other assets needed to enable the functional nodes.

To assess the criticality of the physical nodes, an optimization-based approach was developed for a defender to red-team a potential mission attack. The optimization allows the defender to identify critical nodes assuming differing levels of the attacker's resources. A robust optimization model extends this analysis by capturing the uncertainty of the cost and impact of an attack on a physical node. The robust optimization model could be equivalently interpreted as a model capturing the best choices of mitigations by the defender in anticipation of the actions of the attacker. Finally, two different approaches for ranking physical node criticality were proposed to handle situations where the attacker's resources are completely unknown.

After generating a random class of hybrid mission networks, numerical experiments revealed that for a fixed proportion of physical to functional nodes, the optimization runtime is roughly linear in the size of the network, and large networks with up to 10,000 nodes can be solved within minutes.

References

Ahuja RK, Magnanti TL, Orlin JB (1993) Network flows: theory, algorithms, and applications. Prentice Hall, Upper Saddle River, NJ

Albert R, Albert I, Nakarado GL (2004) Structural vulnerability of the North American power grid. Phys Rev E 69(2):025103

Bell MGH, Kanturska U, Schmöcker JD, Fonzone A (2008) Attacker–defender models and road network vulnerability. Philos Trans R Soc A Math Phys Eng Sci 366(1872):1893–1906

Ben-Tal A, Nemirovski A (1999) Robust solutions of uncertain linear programs. Oper Res Lett 25(1):1–13

Bertsimas D, Sim M (2004) The price of robustness. Oper Res 52(1):35–53
Birnbaum ZW (1968) On the importance of different components in a multicomponent system. Technical Report TR-54, Laboratory of Statistical Research, University of Washington, Washington
Brown GG, Carlyle WM, Salmerón J, Wood RK (2006) Defending critical infrastructure. Interfaces 36(6):530—544
Fussell JB (1975) How to hand-calculate system reliability and safety characteristics. IEEE Trans Reliab 24(3):169–174
Garvey PR, Pinto CA (2009) Introduction to functional dependency network analysis. In: Proceedings of the second international symposium on engineering systems. Massachusetts Institute of Technology, Cambridge, MA
Guariniello C, DeLaurentis D (2017) Supporting design via the System Operational Dependency Analysis methodology. Res Eng Des 28(1):53–69
Humphreys C (1985) Zen Buddhism. Mandala Books, London, p 69
Servi LD, Garvey PR (2017) Deriving global criticality conditions from local dependencies using functional dependency network analysis (FDNA). Syst Eng 20(4):297–306
Shen Y, Nguyen NP, Xuan Y, Thai MT (2013) On the discovery of critical links and nodes for assessing network vulnerability. IEEE/ACM Trans Netw 21(3):963–973
Soyster AL (1973) Convex programming with set-inclusive constraints and applications to inexact linear programming. Oper Res 21(5):1154–1157
Stoer M, Wagner F (1997) A simple min-cut algorithm. J ACM 44(4):585–591
van der Borst M, Schoonakker H (2001) An overview of PSA importance measures. Reliab Eng Syst Saf 72(3):241–245
Vesely WE (1970) A time-dependent methodology for fault tree evaluation. Nucl Eng Des 13(2):337–360
Wasserman S, Faust K (1994) Social network analysis: methods and applications. Cambridge University, Cambridge
Wood RK (1993) Deterministic network interdiction. Math Comput Model 17(2):1–18

References

Chapter 5
Disrupting Functional Pathways Having Supporting Physical Systems: A BSI Approach

One of the glories of English simplicity is the possibility of using the same word as noun and verb.

Edward Sapir (Rausand and Høyland 1931)

5.1 Introduction

Identifying components most critical to a large complex system or network's successful operation is a fundamental concern. Yet, there is no universally preferred definition of criticality among the many alternatives; the domain of application and the specific need for criticality analysis will drive the best choice. As noted in Chap. 3, the literature separates a critical node analysis into two distinct viewpoints, based on whether or not the network topology is the sole consideration for critical node identification. Examples of supplementing network topology with a functional dependency relationship between the nodes include (Garvey and Pinto 2009; Garvey et al. 2014; Servi and Garvey 2017; Guariniello and DeLaurentis 2017).

Critical node analysis using topology alone falls into the categories of graph analysis (e.g., examining the degree of centrality of the nodes) (Wasserman and Faust 1994), pairwise connectivity (Shen et al. 2013; Ventresca and Aleman 2015), cutsets (Ahuja et al. 1993; Stoer and Wagner 1997), global clustering coefficients (Costa et al. 2007), and the shortest paths (Albert et al. 2004). Additionally, an example of supplementing the topology with flow information is Nagurney and Qiang (2008). Alternatively, and it is also the analysis of interest in this chapter, there is a reliability/logic-based viewpoint in which the functionality of the system depends solely on the topology of the nodes, e.g., Rausand and Høyland (2004). Such reliability analysis includes measures such as risk reduction (van der Borst and Schoonakker 2001), Fussell-Vesely importance (Fussell 1975; Meng 2000; Vesely 1970), risk reduction worth, critical importance, risk achievement, risk achievement worth, and Birnbaum Structural Importance (BSI).

This chapter focuses on BSI, in which, as in Chap. 4, the mission under consideration is modeled as a network consisting of a set of physical and functional

L. Servi, *Mission Dependency Network Analysis*, Risk, Systems and Decisions,
https://doi.org/10.1007/978-3-032-11260-6_5

nodes, which include a *START* node, an *END* node, and edges between the nodes. The functional nodes might be viewed as tasks or verbs and the physical nodes as supporting equipment or nouns. The system is said to be operational if there exists a path of operable functional nodes between START and END. BSI was introduced by Z.W. Birnbaum (1968) as a way of assessing individual component importance in the context of a larger system. This chapter expands the missions under consideration to hybrid mission networks by assuming that a functional node is operable if certain specified combinations of physical nodes that support it are operable.

BSI is a well-established system metric of nodal criticality in reliability systems. When applied to a node n, it can be interpreted as the probability that mission will be operational if and only if node n is operational in the face of random instances of failures and non-failures of other nodes.

In the following section, we more formally introduce a hybrid mission network for examining systems using BSI, which expands its applicability. Unfortunately, although BSI is an important measure, it is very burdensome to compute exactly for large systems. Thus, this chapter presents an approach to overcome the computational burden with a fast approximation guaranteed to be within specified error bounds.

This chapter is organized as follows: After presenting the traditional representation of a mission network, a broader new class of system representation, *hybrid mission network*, is introduced in Sect. 5.1.1, which broadens the applicability of the BSI analysis by now having both functional and physical nodes. Section 5.2 presents the analysis of the traditional network representation, followed by that of the hybrid network representation. Section 5.3 overcomes the computational burden of the algorithms in Sect. 5.2 for large systems by presenting a very fast approximation algorithm which is guaranteed to be within a specified error bound, ϵ, with a known probability. The approximation algorithm typically requires orders of magnitude less computation than the exact $O(2^N)$ algorithm, as it is $O(\epsilon^{-2}N^2)$, where N is the number of physical nodes in the network. Section 5.4 presents examples that are used to illustrate the computations and their timing. Finally, Sect. 5.5 summarizes the chapter.

5.1.1 Traditional and Hybrid Network Representations

Traditionally, BSI is applied to the operability of a system consisting of physical nodes, each of which, independent of each other, may or may not be operable. BSI traditionally views a mission as operable if there is a path from a *START* node to an *END* node where all nodes are operable. In this chapter, we introduce a hybrid network representation that can broaden the use of BSI, as well as its applicability by considering two different types of simple nodes: physical nodes and functional nodes. The traditional network could be viewed as having composite nodes with the properties of *both* functional and physical nodes. A hybrid network might represent

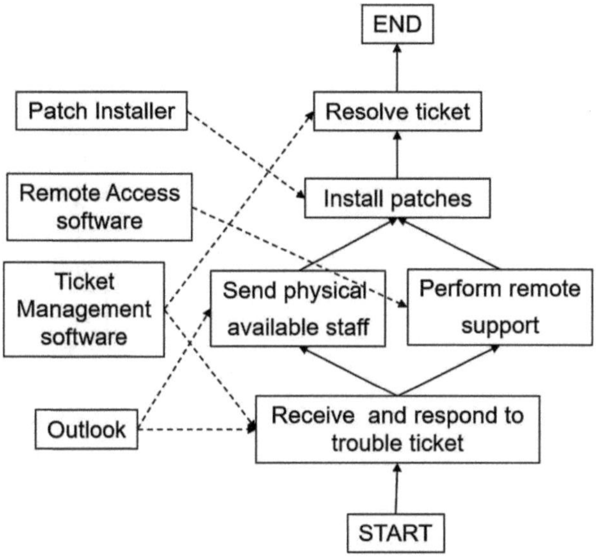

Fig. 5.1 A hybrid simplified representation of an IT Help support activity

a mission made up of functional nodes, which are in turn enabled by physical nodes. The operability of each physical node is independent of other physical nodes; the operability of a functional node depends on whether a prespecified combination of supporting physical nodes is operable. For both traditional and hybrid representations, we say that the mission is operable if there exists a path from the *START* node to the *END* node consisting entirely of operable functional nodes.

Figure 5.1 illustrates an example of a hybrid network that models a simplified process to perform IT support. The functional nodes describe required actions, e.g., "receive and respond to trouble ticket" or "send physical available staff." The operability of the physical nodes is required to accomplish these functions. For example, the physical node "Patch Installer" is required for functional node "Install patches." The success of the IT support activity depends on the physical nodes (in this example, various software) being operable as they enable the functional nodes. Therefore, it is the physical nodes, not the functional ones, that should be examined for their criticality.

Depending on the context, this hybrid network representation of simple physical and functional nodes may offer a more natural and transparent representation than the traditional representation having composite nodes. For instance, a network of functional nodes may represent a set of ordered functions that need to be successful to complete a mission or an activity. Performing all the required functions necessitates the operability of a combination of certain physical nodes. Therefore, the equivalent hybrid network representations could, in principle, be represented exactly using a complex traditional network representation. However, the transparency of the model would be lost.

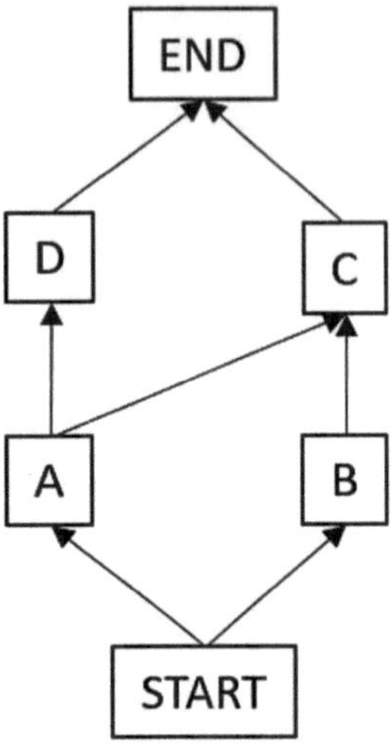

Fig. 5.2 A traditional network representation

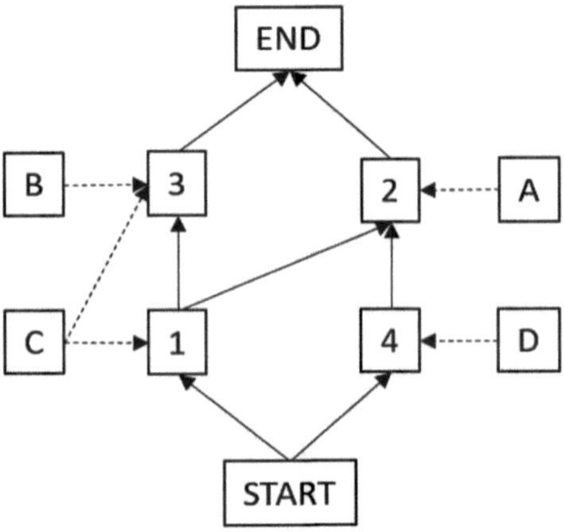

Fig. 5.3 A hybrid network representation

Figures 5.2 and 5.3 illustrate the same system using the traditional and hybrid network representations, respectively, and demonstrate the equivalency relationship between them. In Fig. 5.2, nodes A, B, C, and D correspond to composite nodes. They are physical nodes in the sense that each may or may not be operable, independent of the others. They are also functional in the sense that there are paths of such nodes, whose components must all be operable for the system to be operable. Figure 5.2 shows three paths for success: $START \rightarrow A \rightarrow D \rightarrow END$, $START \rightarrow A \rightarrow C \rightarrow END$, and $START \rightarrow B \rightarrow C \rightarrow END$. This implies success if (1) A and D are both operable, (2) A and C are both operable, or (3) B and C are both operable.

Figure 5.3 illustrates a hybrid network representation, where nodes 1, 2, 3, and 4 represent the functional nodes, and nodes A, B, C, and D represent the physical nodes. Each physical node may or may not be operable, independent of the others.

Each functional node is either operable or inoperable, based on the status of the physical nodes connected to it. For this example, functional node 1 is operable if and only if physical node C is operable; functional node 2 is operable if and only if physical node A is operable; functional node 3 is operable if and only if physical nodes B and C are both operable; and functional node 4 is operable if and only if physical node D is operable. There are three alternative paths for a successful activity: *START* $\rightarrow$ 1 $\rightarrow$ 2 $\rightarrow$ *END*, *START* $\rightarrow$ 1 $\rightarrow$ 3 $\rightarrow$ *END*, and *START* $\rightarrow$ 4 $\rightarrow$ 2 $\rightarrow$ *END*. This implies the activity will succeed if (1) A and C are both operable, (2) B and C are both operable, or (3) A and D are both operable. Hence the mission in Fig. 5.3 will succeed if and only if the mission in Fig. 5.2 succeeds.

From a practical point of view, the hybrid network representation is qualitatively different from the traditional network representation, is more versatile, and is more natural to use. However, as foreshadowed in the discussion of Figs. 5.2 and 5.3, the two representations can be made equivalent. A hybrid network representation can be converted into a number of derived sets of physical nodes in which all nodes must be operable for the mission to succeed. If each such set corresponded to a path from *START* to *END* in a traditional network representation, a mission would succeed under precisely the same logical conditions as the hybrid representation. As such, any hybrid network representation can be modeled as an equivalent traditional network representation. In the example with Figs. 5.2 and 5.3, it was assumed that the operability of a functional node depended on the AND of a set of physical nodes. This assumption could be generalized to any logical relationship such that a physical node changing from inoperable to operable cannot cause a functional node to change from operable to inoperable.

One might also note that the hybrid network representation can be used to represent *any* function consisting of combinations of AND and OR functions. One, of many, realization of an arbitrary function would use the topology of the functional nodes to capture the OR functions and the structure of the supporting physical nodes to capture the AND functions.

While time is not explicitly incorporated in traditional and hybrid network representations, it can be captured implicitly. For example, in Fig. 5.3, functional node 3 can only be performed after functional node 1. Also, the representations assume that a physical node failure *immediately* impacts functional node operabilities and consequently the system's operability. When a physical node failure has a lagged effect on a functional node or the system operability, this temporal effect can be captured by redefining a node to represent a specific node's operability *at a specified time*. The network representation would then need parts or the entire original network duplicated for each time instant of interest. If there are multiple nodes corresponding to the END node at different instants of time, they must be aggregated together into a single final node using the AND function, OR function, or some other function. With this representation modification, the static analysis of this chapter will still be applicable and now captures the temporal effects.

5.1.2 Definitions and Basic Relationships

The following definitions and notation will be used in this chapter:

- Physical node: a node that can be either operable or inoperable and is independent of all other physical nodes. A physical node might become inoperable due to external independent causes. A physical node might also be a functional node.
- Functional node: a node that is an element of a path from START node to END node. Each functional node is operable or inoperable based on the status of its supporting physical nodes.
- Network: a set of nodes consisting of a START node, an END node, physical nodes that might also be functional nodes, functional nodes that are not physical nodes, a set of directed links between the functional nodes, and, if some physical nodes are not also functional nodes, a supporting combination of physical nodes for each functional node.
- Path: a noncyclic, directed set of connected functional nodes, which is considered operable if and only if every functional node in the set is operable. The START node and END node are functional nodes.
- $\mathcal{N}$ is the set of physical nodes. $N = |\mathcal{N}|$.
- $\mathcal{N}_f$ is the set of functional nodes. $N_f = |\mathcal{N}_f|$. For the case of a traditional network $\mathcal{N} = \mathcal{N}_f$.
- $\mathcal{A}_f$ is the set of arcs between functional nodes. $A_f = |\mathcal{A}_f|$.
- $\mathcal{F}(n) = \{m \in \mathcal{N}_f : (n, m) \in \mathcal{A}_f\}$ is the set of functional nodes that directly follow functional node n in the network.
- $\mathcal{H}$ is the set of all paths of functional nodes from *START* to *END*.
- d_n is the number of distinct subsets of physical nodes such that, if all are operable, functional node n is operable.
- $x = (x_1, x_2, \ldots, x_N)$ is a binary *state vector* capturing the state of each physical node. $x_n = 1$ indicates that node n is operable and $x_n = 0$ indicates inoperability.
- $x^{(1,n)} = (x_1, \ldots, x_{n-1}, 1, x_{n+1}, \ldots, x_N)$. That is, $x^{(1,n)}$ is the vector formed by taking x and replacing the nth component with a 1.
- $x^{(0,n)} = (x_1, \ldots, x_{n-1}, 0, x_{n+1}, \ldots, x_N)$. That is, $x^{(0,n)}$ is the vector formed by taking x and replacing the nth component with a 0.
- $y(x) = (y_1(x), y_2(x), \ldots, y_{N_f}(x))$ is a binary state vector capturing the state of each functional node. $y_n(x) = 1$ indicates that functional node n is operable given x and $y_n(x) = 0$ indicates inoperability.
- $C_k(n) \subset \mathcal{N}$ is the kth set of physical nodes that, if all operable, will make the functional node n operable, for at least one $k \in \{1, \ldots, d_n\}$. We refer to these

subsets as *supporting combinations*. For a traditional network representation, where each physical node is also the functional node which it supports, $d_n = 1$, with $C_1(n) = n$. For a hybrid network representation where a functional node does not require any physical nodes to be operable, we use the convention $d_n = 1$, with $C_1(n) = \emptyset$, which implies $y_n(x) = 1$ for all x.

- $\phi : \{0, 1\}^N \mapsto \{0, 1\}$ is the *structure function* for the system. This indicates that if a state vector x results in system failure, $\phi(x) = 0$; otherwise, $\phi(x) = 1$.
- *Critical path vector (CPV)* with respect to node n: This is a property of x such that there is an operable path from START to END that includes $x_n = 1$. However, by switching x_n to 0 without modifying the state of any of the other $N - 1$ physical nodes, the path loses operability. Mathematically,[1] this can be expressed as follows: x is a critical path vector for component n if $\phi(x^{(1,n)}) = 1$ and $\phi(x^{(0,n)}) = 0$ which implies $\phi(x^{(1,n)}) - \phi(x^{(0,n)}) = 1$.
- *Birnbaum Structural Importance (BSI)*, B_n for node n: This is the probability that a given state vector is a critical path vector for node n, assuming all state vectors are equally likely to occur (Birnbaum 1968).
- *Resilience metric* (R): the fraction of state vectors, x, that result in the system working, i.e., $R = \sum_x \phi(x)/\sum_x 1$. This is a number between 0 and 1. The closer R is to 1, the more likely the mission will be operable if the state of each system is random, and the more resilient the system is.

By definition, the structure function $\phi(x) = 1$ if and only if for at least one path, $p \in \mathcal{H}$, all functional nodes on p are operable, i.e.,

$$\phi(x) = 1 - \prod_{p \in \mathcal{H}} \left(1 - \prod_{n \in p} y_n(x)\right). \tag{5.1}$$

A functional node n is operable, i.e., $y_n(x) = 1$, if there is at least one supporting combination of n that consists of exclusively operable physical nodes, i.e.,

$$y_n(x) = 1 - \prod_{k=1}^{d_n} \left(1 - \prod_{m \in C_k(n)} x_m\right). \tag{5.2}$$

In the case of a traditional network representation, $d_n = 1$ and $C_1(n) = n$, which implies $y_n(x) = x_n$, Eq. (5.1) simplifies to

$$\phi(x) = 1 - \prod_{p \in \mathcal{H}} \left(1 - \prod_{n \in p} x_n\right). \tag{5.3}$$

---

[1] Inherent in this definition is the assumption that the system is *coherent*. This means that the structure function is monotone in x (i.e., changing a component from operable to inoperable should never cause system failure). The majority of this work will deal with network connectivity, which is one type of coherent system.

Finally, by definition (Birnbaum 1968), B_n is the fraction of combinations of x such that $\phi(x^{(1,n)}) = 1$ and $\phi(x^{(0,n)}) = 0$, so

$$B_n = 2^{-N} \sum_x \left[\phi(x^{(1,n)}) - \phi(x^{(0,n)}) \right]. \tag{5.4}$$

BSI can be used to understand the relative importance of node operability in a network. As shown in the Appendix, (5.4) could be interpreted as a linear regression solution. Using the definitions for BSI, the state vector, x, $x^{(1,n)})$, $x^{(0,n)}$, as well as the structure function ϕ, the BSI is given by (5.4).

While (5.4) is the classical definition of BSI, an equivalent form is computationally more useful. Specifically, since the summand in (5.4) does not depend on the value of x_n, identical summand terms in (5.4) appear twice: when $x_n = 0$ and when $x_n = 1$. Hence, (5.4) is equivalent to multiplying the whole expression by two and then summing over the vectors x where either $x_n = 1$ or $x_n = 0$. This simplification is shown below:

$$\begin{aligned} B_n &= 2^{-(N-1)} \left(\sum_{\{x:x_n=1\}} \phi(x) - \sum_{\{x:x_n=0\}} \phi(x) \right) \\ &= 2^{-(N-1)} \left(\sum_x x_n \phi(x) - \sum_x (1 - x_n)\phi(x) \right) \\ &= 2^{-(N-1)} \sum_x \phi(x)(2x_n - 1). \end{aligned} \tag{5.5}$$

Computing the BSI for a given node requires evaluating $\phi(x)$ for all 2^N possible state vectors x. However, once the 2^N evaluations of ϕ have been performed, there is enough information to compute B_n for *all* $n \in \mathcal{N}$. This can be done by using the following equation:

$$\vec{B} = 2^{-(N-1)} \sum_x \phi(x)(2x - \vec{1}), \tag{5.6}$$

where $\vec{B}$ is a vector with components $\vec{B}_n = B_n$, and $\vec{1}$ represents the vector of all ones.

Equation (5.6) has the advantage that it does not require saving ϕ values for the 2^N state vectors, since the BSI vector instead can be incrementally updated with $\phi(x)$ for each state vector x and not saved. Equation (5.6) also allows concurrent computation of the BSI vector for every state x. Note, also, that the derivation of (5.6), and in fact most of this chapter, can be easily generalized for the case of all state vectors not being equally likely.

5.2 Efficiently Computing the Structure Function, $\phi(x)$

As indicated in (5.6), an essential component in computing the Birnbaum Structural Index is the ability to efficiently compute the structure function $\phi(x)$. Although (5.1) provides a closed form for the structure function, using it in practice would require knowing or computing all of the network paths a priori. Even when this knowledge is known, it may be memory-intensive to store the full set of such paths. Instead, this section presents algorithms that computes $\phi(s)$ using only local node information. This is done first for the traditional network case and then for the hybrid network case.

5.2.1 The Traditional Network Case

The key idea behind making the computation of (5.3) more memory-efficient for a traditional network is to compute $\phi(x)$ by traversing the operable portion of the network beginning at the *START* node and working forward to determine if any *START*-to-*END* path is operable. We will rely on the set $\mathcal{F}(n)$, formally defined in Sect. 5.1.2, which contains out-neighbors of node n. This is accomplished by performing a depth-first search (DFS) on the network, beginning with the *START* node. The search will terminate if it reaches the *END* node at any point and will return $\phi(x) = 1$. If the search traverses the entire subgraph without finding the *END* node, the *END* node must not be reachable from *START* with operable functional nodes. As such, the algorithm will return $\phi(x) = 0$.

To implement this algorithm, we keep track of two sets of nodes. The first set, $Uninterested$, contains the set of nodes which either the search has already examined or which are inoperable. As the network may contain multiple paths between two nodes, membership in this set ensures that no node is visited more than once. We initialize the set *Uninterested* with the inoperable nodes $\{n \in \mathcal{N} : x_n = 0\}$, because they should not be considered in the search.

The second set, *Boundary*, contains the boundaries of the search algorithm at each step. If a node m is in *Boundary*, there is an operable *START* $\rightarrow m$ path, and the algorithm has not ruled out the possibility that there could be an operable $m \rightarrow$ *END* path. The algorithm terminates if the *END* node ever enters *Boundary* (implying there is a *START* $\rightarrow$ *END* path, so $\phi(x) = 1$) or if *Boundary* becomes empty (implying no such path so $\phi(x) = 0$).

To create the DFS behavior, this set is implemented as a stack (Papadimitriou and Steiglitz 1998). The algorithm uses a binary flag $EndReachable$ to indicate if the *END* node has entered the set *Boundary*. When the algorithm terminates, the value of ϕ will equal the $EndReachable$ flag. Complete details of computing the structure function for a given state vector are shown in Algorithm 4. In the worst

Algorithm 4 Computing the value of the structure function ϕ for a given instance x

Input: $x \in \{0, 1\}^N, \mathcal{F}$
Output: $\phi(x) \in \{0, 1\}$

Uninterested $\leftarrow \{n \in \mathcal{N} : x_n = 0\}$
Boundary $\leftarrow$ {*START*}
EndReachable $\leftarrow 0$
while Boundary $\neq \emptyset$ and EndReachable $= 0$ **do**
  $n \leftarrow$ next element in *Boundary*
  Boundary $\leftarrow Boundary \cup \{m \in \mathcal{F}(n) : m \notin Uninterested\}$
  if END $\in Boundary$ **then**
    $EndReachable \leftarrow 1$
  end if
  Remove n from *Boundary*
  Uninterested $\leftarrow$ *Uninterested* $\cup \{n\}$
end while
  return $\phi(x) \leftarrow EndReachable$

case, this algorithm runs like a standard DFS, which is $O(A)$[2] (Papadimitriou and Steiglitz 1998). However, this search will only explore the nodes that remain in the connected component with the *START* node and will stop if it ever reaches the *END* node. Both of these details will increase runtime efficiency in practice.

One approach to potentially gain some additional computational savings is to have Algorithm 4 return the operable path that is discovered if $\phi(x) = 1$. In this case, if $\phi(x) = 1$ and node n is not an element of the discovered path, $\phi(x^{(0,n)})$ must still equal one. However, in practice, the probability that $\phi(x)$ is zero is typically much higher than that of $\phi(x)$ being one. Therefore, the benefits of avoiding the computation of $\phi(x^{(0,n)})$ in these cases would need to be weighed against the costs of factoring this type of check into the algorithms.

5.2.2 The Hybrid Network Case

The algorithm in Sect. 5.2.1 could be adapted to a hybrid network representation by adding a simple preprocessing step. For each functional node n, compute the operability $y_n(x)$ based on the given state vector of the physical nodes, x. Then, Algorithm 4 could be run directly on the functional network, with functional node states given by the vector $y(x)$. However, a more efficient approach would be to perform the graph search directly on the network of functional nodes, only computing the operability score when a functional node is initially encountered.

[2] DFS is often reported as O(N+A). However, for our application, the number of arcs A will dominate the number of nodes N, because we are considering connected networks with at least two *START* – *END* paths.

Algorithm 5 Computing the value of the structure function ϕ for a hybrid network

```
Input: x ∈ {0, 1}^{N_p}, 𝓕, C
Output: φ(x) ∈ {0, 1}
Uninterested ← ∅
Boundary ← {START}
EndReachable ← 0
while Boundary ≠ ∅ and EndReachable = 0 do
    n ← next element of Boundary
    k ← 1
    comboWorking ← 0
    while comboWorking = 0 and k ≤ d_n do
        if x_j = 1 for all j ∈ C_k(n) or C_k(n) = ∅ then
            comboWorking ← 1
            Boundary ← Boundary ∪ {m ∈ 𝓕(n) : m ∉ Uninterested}
        end if
        k ← k + 1
    end while
    Remove n from Boundary
    Uninterested ← Uninterested ∪ {n}
    if END ∈ Boundary then
        EndReachable ← 1
    end if
end while
    return φ(x) ← EndReachable
```

This reduces the overall computational time by ignoring the nodes that will not be reached in the graph search.

This algorithm operates in a very similar fashion to Algorithm 1. Here, the graph search is conducted on the network of functional nodes, beginning with the *START* node. At each step, we choose a functional node n from the $Boundary$ set and seek to determine its operability. Functional node n is operable if there exists a k such that $x_j = 1$ for all $j \in C_k(n)$. The algorithm keeps track of this for each node with a binary flag called *comboWorking*. If this flag changes to True for node n, we know that n is operable and that there is a *START* $\rightarrow$ n path. If a functional node n is determined to be operable, the following nodes $m \in \mathcal{F}(n)$ can join $Boundary$ and be examined later. If node n is inoperable, we add it to the set $Uninterested$. In both cases, n is removed from the $Boundary$. The algorithm will terminate if the *END* node enters the $Boundary$ set or once $Boundary$ becomes empty. This can be seen in Algorithm 5.

To determine the runtime of Algorithm 5, note that the primary difference between Algorithms 4 and 5 is the requirement to find if a combination of physical nodes is operable that supports functional node n. In the worst case, this process will take $\sum_k |C_k(n)|$ computations, for which a very conservative bound is $N \max_i(d_i)$. The other element inside this loop is to add members of $\mathcal{F}(n)$ to the boundary set, which will occur at most once for each arc (n, m), where $m \in \mathcal{F}(n)$. Summing the size of this set over all functional nodes n will yield the size of the complete arc set, A_f. The remainder of the computations is constant and done for each functional node n. Therefore, the worst-case runtime for this algorithm would be

$$\sum_{n \in \mathcal{N}_f} \left(O(1) + |\mathcal{F}(n)| + \sum_{k=1}^{d_n} |C_k(n)| \right) = O(A_f + |\mathcal{N}_f||\mathcal{N}|). \tag{5.7}$$

5.3 Efficiently Approximating and the BSI

The methods of Sect. 5.2 permit one to efficiently compute the structure function, $\phi(x)$, and hence the BSI using (5.6). Computation of BSI can be radically sped up if a variation of (5.6) is employed to estimate the BSI using only a sample of the space of all state vectors, x. To do this rigorously requires that bounds are found that characterize the accuracy of the resulting estimate. This is done in Sect. 5.3.1. Section 5.3.2 presents the details for estimating the BSI. In some cases, it is less important to characterize the criticality of *all* the physical nodes than to identify the top m critical physical nodes. Section 5.3.3 addresses this objective. Finally, Sect. 5.3.4 presents an approximate approach to characterizing the cutsets of a network quickly. Cutsets are of interest as they can sometimes be viewed as an alternative view of the criticality of a node.

5.3.1 *Finding Rigorous Probabilistic Bounds for the BSI*

The computational bottleneck for finding the Birnbaum Structural Importance measure, B_n, exactly stems from the need to examine 2^N combinations of state vectors, as each of the N physical nodes may be either operable or inoperable. When N is large, it may not be feasible to compute the BSI measure exactly. In such cases, one could, instead, approximate the BSI by only sampling from a subset of possible state vectors, $\mathcal{X}$. It will be shown that sample sizes that are many orders of magnitude fewer states than 2^N would still guarantee that the resulting approximation of B_n is within an arbitrarily small error, ϵ, with a specified probability. Specifically, since B_n is the probability that a random state vector is a critical path vector for node n, B_n can be estimated by randomly selecting only the state vectors $\mathcal{X}$ and estimating BSI based only on these samples, $\mathbf{B}_n(\mathcal{X})$.

The random variable, $\mathbf{B}_n(\mathcal{X})$, is the average of $|\mathcal{X}|$ Bernoulli trials, each trial having a probability of being a critical path vector with respect to that trial of node n of B_n. As such, $\mathbf{B}_n(\mathcal{X})$ has a mean of B_n and a standard deviation given by

$$\sigma\left(\mathbf{B}_n(\mathcal{X})\right) = \sqrt{\frac{B_n(1 - B_n)}{|\mathcal{X}|}}. \tag{5.8}$$

From Chebyshev's Inequality (Bienaymé 1853; Saw et al. 1984; Tchebychev 1867), for any $t > 0$,

$$Pr\left[\left|\mathbf{B}_n(\mathcal{X}) - B_n\right| > t\,\sigma(\mathbf{B}_n(\mathcal{X}))\right] \leq \frac{1}{t^2}. \tag{5.9}$$

From (5.9), to guarantee a 99% confidence interval with a width of ϵ, one can set

$$\epsilon = t\,\sigma(\mathbf{B}_n(X)) \tag{5.10}$$

and set the right-hand side of (5.9) to 1%, i.e., set $t = 10$. From (5.9) and (5.10), the number of state vectors needed to compute a sufficient approximation of BSI measure is given by

$$|\mathcal{X}| = \frac{B_n(1 - B_n)t^2}{\epsilon^2}. \tag{5.11}$$

Since, for all B_n between 0 and 1, $B_n(1 - B_n) \leq 1/4$, an upper bound for Eq. (5.11) independent of B_n is given by

$$|\mathcal{X}| = \frac{t^2}{4\epsilon^2}. \tag{5.12}$$

For large $|\mathcal{X}|$, the binomial distribution can quite accurately be approximated by a normal distribution. Hence, $t = 2.94$ rather than $t = 10$ is all that is practically needed for a 99% confidence interval of width ϵ. This upper bound is $|\mathcal{X}| \approx \frac{2.94^2}{4(.05^2)}$ which, for $\epsilon = 0.05$, is approximately 900. Therefore, as a rule of thumb, a 99% confidence interval with a width of 0.05 can be provided by selecting $|X| > 900$, which further takes advantage of the normal approximation to the binomial distribution. This difference creates many orders of magnitude of computational savings.

5.3.2 Approximating the BSI

Algorithm 6 below shows how to create an approximate BSI (denoted by $\widehat{B}$), using a sample of state vectors $\mathcal{X}$. In the case where $\mathcal{X} = \{0, 1\}^N$, the algorithm will output the exact BSI. The main portion of the algorithm iterates through each state vector $x \in \mathcal{X}$ and determines if this constitutes a critical path vector for any node n. If so, the component $\widehat{B}_n$ is incremented by one over $|\mathcal{X}|$. When the algorithm ends, the resulting estimate $\widehat{B}_n$ gives the fraction of randomly drawn vectors $x \in \mathcal{X}$ that are a critical path vector for node n. This serves as an estimate for B_n based on the sample $\mathcal{X}$.

The specific mathematics underlining Algorithm 6 is as follows: Motivated by (5.4) and (5.13) follows. A term in the summation of (5.13) can equal one only if $\phi(x^{(1,n)}) = 1$ and $\phi(x^{(0,n)}) = 0$. There are two possibilities:

- $\phi(x) = 1$: Here, if $x_n = 1$ and $\phi(x^{(0,n)}) = 0$, it must be the case that $\phi(x^{(1,n)}) = 1$ and $\phi(x^{(0,n)}) = 0$.
- $\phi(x) = 0$: Here, if $x_n = 0$ and $\phi(x^{(1,n)}) = 1$, it must be the case that $\phi(x^{(1,n)}) = 1$ and $\phi(x^{(0,n)}) = 0$.

Hence, (5.14) and Algorithm 6 follow.

$$\widehat{B}_n = \frac{1}{|\mathcal{X}|}\sum_{x\in\mathcal{X}}[\phi(x^{(1,n)}) - \phi(x^{(0,n)})] \tag{5.13}$$

$$= \frac{\left|\{x : x \in \mathcal{X}, \phi(x)=1, x_n=1, \phi(x^{(0,n)})=0\}\right| + \left|\{x : x \in \mathcal{X}, \phi(x) = 0, x_n = 0, \phi(x^{(1,n)}) = 1\}\right|}{|\mathcal{X}|}. \tag{5.14}$$

Algorithm 6 Approximating the BSI with a sample of state vectors, X

```
Input: 𝒩, ℱ, 𝒳 ⊂ {0, 1}^N
Output: B̂, an approximation to B⃗
Initialize B̂ = 0⃗
for x ∈ X do
    if φ(x) = 1 then
        for n ∈ 𝒩 do
            if x_n = 1 and φ(x^(0,n)) = 0 then
                B̂_n ← B̂_n + 1/|X|
            end if
        end for
    else
        for n ∈ 𝒩 do
            if x_n = 0 and φ(x^(1,n)) = 1 then
                B̂_n ← B̂_n + 1/|X|
            end if
        end for
    end if
end for
  return B̂
```

Algorithm 6 requires that the structure function $\phi(x)$ be evaluated approximately $|\mathcal{N}||\mathcal{X}|/2$ times, but no more than $|\mathcal{N}||\mathcal{X}|$. Therefore, given the worst-case runtime of Algorithm 5, for a (1-α)-confidence interval of width ϵ, the BSI approximation is $O(|\mathcal{N}|A/(\epsilon^2\alpha))$.

This algorithm provides a template for counting the critical path vectors in a set of state vectors. Depending on the use case, it may be possible to gain some computational savings by modifying certain details. For example, if the network is fragile (i.e., the probability that $\phi(x) = 1$ is low), it would be advantageous to ignore the *else* block, provided that the sample size $|\mathcal{X}|$ has been increased accordingly. In such a scenario, it is necessary that the final step of the algorithm divides $\widehat{B}_n$ by the count of vectors $x \in \mathcal{X}$ with $x_n = 1$.

5.3.3 Identifying the Physical Nodes with the Top m BSI Scores

In Sect. 5.3.2, an approximation algorithm was developed to quickly compute B_n. In some cases, instead of seeking values of B_n, one only seeks to identify the physical nodes with the top m BSI scores, as these may be considered the most critical nodes. For this goal, a simple adaptive variation of the approximate algorithm is possible and useful. The basic approach is to first use (5.11) to estimate the number of instances, $|\mathcal{X}_1|$, needed to approximate each B_n. If the values of $\widehat{B}_n$ for two different top values of n are too close to each other, there is a minimal value in using more instances to resolve the differences. If the values differ significantly, again, there is a minimal value in using more instances as the ordering of the nodes is clear. However, between these two possibilities, using more instances will be of value. Specifically, additional instances are sought if $Max[\phi_1^L \widehat{B}_n, \phi_2^L] \leq \widehat{B}_n - \widehat{B}_{n-1} \leq Max[\phi_1^U \widehat{B}_n, \phi_2^U]$, where, for some applications, $\phi_1^L = 0.1$, $\phi_2^L = 0.02$, $\phi_1^U = 0.30$, and $\phi_2^U = 0.05$ are good initial default values. When more instances are needed to resolve the ordering, based on the values found and how close the top scores are to each other, Eq. (5.11) can be used to compute an additional number of instances required to identify the top scores, $|\mathcal{X}_2|$. This adaptive process may be repeated multiple times. As noted, a good initial value for $|\mathcal{X}_1|$ is 900.

The above approach requires that a new estimate of B_n is found based on multiple iterations of the above approach. If $|\mathcal{X}_1|$ instances are observed to compute $\widehat{B}_n^1$ and an additional $|\mathcal{X}_2|$ instances chosen to be observed to compute $\widehat{B}_n^2$, it is a simple matter to verify that the combined observations lead to the following $\widehat{B}_n$, if all instances are equally probable:

$$\widehat{B}_n = \frac{|\mathcal{X}_1|\widehat{B}_n^1 + |\mathcal{X}_2|\widehat{B}_n^2}{|\mathcal{X}_1| + |\mathcal{X}_2|}. \tag{5.15}$$

Below is a formal description of two iterations of the algorithm.

5.3.4 Birnbaum Cutsets

A *cutset* is defined as a set of nodes in a graph with the property that all paths from the START node to the END node must pass through at least one node of this set. A *minimum cutset* is a cutset of a graph with the property that no other cutset has

Algorithm 7 Finding the physical nodes with the top m BSI scores

Input: m, $|\mathcal{X}_1|$, $|\mathcal{X}_2|$
Output: An ordering of the top m nodes
Compute $\widehat{B}_n$ for all nodes using X_1 number of instances and sort set of nodes with Birnbaum measures $\widehat{B}_1^1 \geq \widehat{B}_2^1 \geq \cdots$
for nodes $n = 2, 3, \ldots m$ **do**
  if $Max[\phi_1^L \widehat{B}_n^1, \phi_2^L] \leq \widehat{B}_n^1 - \widehat{B}_{n-1}^1 \leq Max[\phi_1^U \widehat{B}_n^1, \phi_2^U]$ **then**
    Use equation (5.11) to compute an additional number of desired instances, $\mathcal{X}_2|$.
    Aggregate the computations of $\widehat{B}_n^1$ and $\widehat{B}_n^2$ using (5.15) if instances are equally probable.
  end if
end for
  return Top m nodes

fewer elements. A prominent algorithm for computing the minimum cutset (Stoer and Wagner 1997) requires $O(NE + N^2 \log N)$, where N is the number of nodes and E is the number of edges in the network. cutsets and minimum cutsets are sometimes used to identify critical nodes in a graph. For some applications, it is not necessary to find a minimum cutset but only a cutset with a small number of nodes.

This section describes a fast approach for doing so. The basic approach leverages the observation that nodes in subsets of cutsets with low cardinality tend to have high BSI values. It applies a greedy algorithm using BSI measures to compute what might be called a *Birnbaum cutset* to systems whose nodes are both physical and functional nodes. This approach is much faster than a traditional cutset algorithm that yields sets of interest at the expense of not guaranteeing to find the true minimum cutset.

Algorithm 8 finds the Birnbaum cutset and can be executed using either the exact or approximate Birnbaum algorithms. Define C as the Birnbaum cutset. Initially, set C equal to the null set, and set $\mathcal{N}$ as the network.

Algorithm 8 Finding Birnbaum cutsets

Input: $\mathcal{N}$
Output: C
while There exists at least one path from START to END in the network **do**
  for each node $n \in \mathcal{N}$ **do**
    Compute $\widehat{B}_n$
  end for
  $n^* = argmax_n \widehat{B}_n$
  $C = C \cup \{n^*\}$
  Delete n^* from the network along with all links and paths which contain n^*
end while
  return C

5.4 Numerical Examples

The algorithms introduced in Sect. 5.3 are illustrated on a 7-node directed network, an 11-node undirected network, a 21-node directed network, and a 112-node directed network. For simplicity, in all examples, it is assumed that the nodes are both physical and functional. In each case, $\vec{B}$ was calculated by iterating through either all or a subset of all possible state vectors, x, computing $\phi(x)$ with Algorithm 4, and updating the summation in (5.6). Runtime results of the examples are summarized in Table 5.1 for the exact BSI algorithm and the approximate BSI algorithm after setting $\epsilon = .05$. Note in Table 5.1 that for Examples 1 and 2 (i.e., with 7 nodes and 11 nodes, respectively), the number of instances required for the exact algorithm is so small that the approximate algorithm cannot offer any improvement. Also, for Example 4 (i.e., with 112 nodes), it is infeasible to compute the exact algorithm, so the timing was estimated by the counting of operations using (5.7).

Example 1: A Seven-node Directed Network To illustrate the concept of a BSI measure and the critical path vectors, consider the network in Fig. 5.4 consisting of the START node, the END node, a node of interest, node 5, and four other nodes, nodes 1, 2, 3, and 4. There exist four paths between *START* and *END*: (1) *START* →

Table 5.1 Runtime comparisons

Example	Number of nodes	Exact algorithm (sec)	Approximate algorithm (sec)
1	7	0.004	–
2	11	0.024	–
3	21	10.6	0.15
4	112	10^{39}	30.4

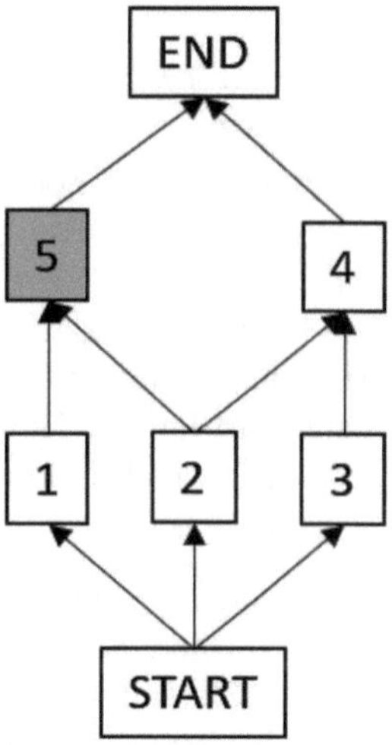

Fig. 5.4 Example 1: a 7-node network

Table 5.2 Instances and critical path vector, CPV, indicators for node 5

Node 1	0	0	0	0	0	0	0	0	1	1	1	1	1	1	1	1
Node 2	0	0	0	0	1	1	1	1	0	0	0	0	1	1	1	1
Node 3	0	0	1	1	0	0	1	1	0	0	1	1	0	0	1	1
Node 4	0	1	0	1	0	1	0	1	0	1	0	1	0	1	0	1
CPV	No	No	No	No	Yes	No	Yes	No	Yes	Yes	Yes	No	Yes	No	Yes	No

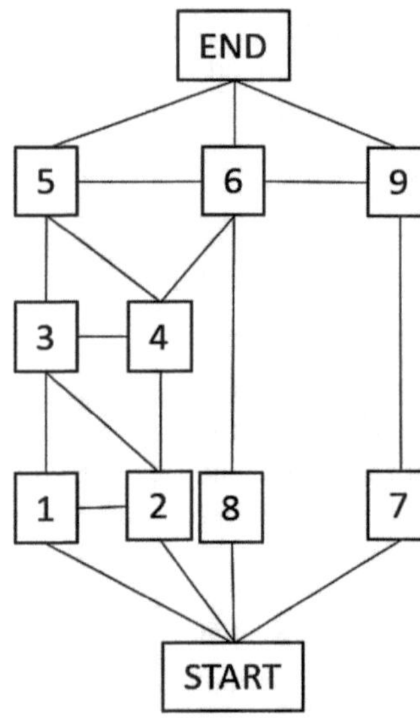

Fig. 5.5 Example 2: an 11-node network

Table 5.3 The exact Birnbaum Structural Importance (BSI) for each node

Node index	1	2	3	4	5	6	7	8	9
BSI value	0.0820	0.1758	0.1523	0.1289	0.2227	0.3398	0.2617	0.2227	0.2617

$1 \rightarrow 5 \rightarrow END$, (2) $START \rightarrow 2 \rightarrow 5 \rightarrow END$, (3) $START \rightarrow 2 \rightarrow 4 \rightarrow END$, and (4) $START \rightarrow 3 \rightarrow 4 \rightarrow END$. As indicated in Table 5.2, there are 16 different combinations for nodes 1, 2, 3, and 4 to be operable or not, denoted by a 1 or 0, respectively. Only 7 of these 16 combinations have the property that there is a path between *START* and *END* with exclusively operable nodes if and only if node 5 is operable, i.e., it is the critical path vectors for node 5. Hence, $B_5 = 7/16$. An implementation of the algorithm can reproduce this result in 4 milliseconds on a standard laptop. The Birnbaum cutset computed using Algorithm 8 is found to be $\{4, 5\}$, which is clearly also the minimum cutset. The resilience metric, from Sect. 5.1.2, is 0.59, reflecting a high redundancy of this network.

Example 2: An 11-node Undirected Network Consider the 11-node network in Fig. 5.5, where all links are undirected. The BSI values are given in Table 5.3 and required 24 milliseconds to compute. The Birnbaum cutset is found to be $\{5, 6, 7\}$, which is also the minimum cutset. The resilience metric is 0.61.

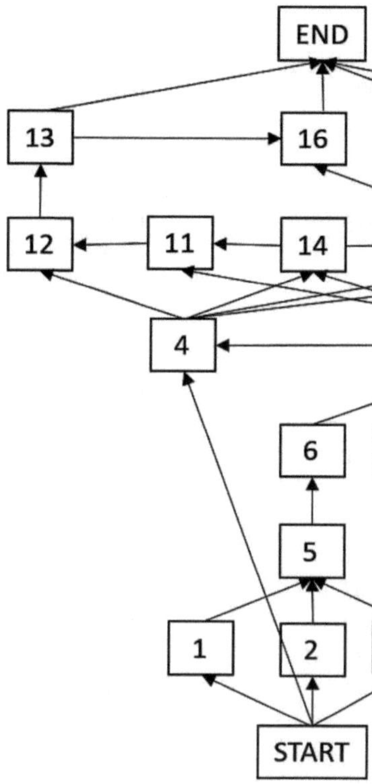

Fig. 5.6 Example 3: a 21-node network

Example 3: A 21-node Directed Network Consider the 21-node network in Fig. 5.6. The BSI measure for all nodes was computed exactly in 10.6 seconds, and the approximation using only 900 instances resulted in 0.15 seconds. It can be seen in Table 5.4 that the approximation was quite accurate despite the two orders of magnitude reduction in computational running time. The Birnbaum cutset, computed in 12 seconds using the algorithm in Sect. 5.3.4, was found to be {5, 6}, which is a minimum cutset—the other minimum cutsets are {5, 7} and {5, 8}. The resilience metric is 0.35. Perhaps this low value reflects the high dependency of the single node 4 being operational.

Example 4: A 112-node Directed Network Consider the 112-node network in Fig. 5.7, which has 16,682 paths. The BSI measures for all nodes were computed approximately in 30.4 seconds. From Table 5.1, one can estimate that the exact calculation would require 10^{39} seconds. Figure 5.8 computes the BSI values with three standard deviations bars. The six nodes with the highest BSI measures, from the algorithm in Sect. 5.3.3, are found to be, in order, nodes 90, 91, 89, 67, 26, and 30, after 49 seconds of computing. The Birnbaum cutset, computed in 130 seconds using the algorithm in Sect. 5.3.4, was found to be {91, 46, 26, 69, 77, 54, 16}. This might have been difficult to find by inspection. The resilience metric is 0.37.

Table 5.4 The exact and approximate Birnbaum Structural Importance (BSI) for each node

Node index	Exact BSI	Approximate BSI	BSI standard deviation
1	0.007	0.007	0.003
2	0.007	0.003	0.002
3	0.007	0.003	0.002
4	0.597	0.593	0.016
5	0.051	0.043	0.007
6	0.051	0.060	0.008
7	0.051	0.051	0.007
8	0	0	0
9	0	0	0
10	0	0	0
11	0.008	0.010	0.003
12	0.125	0.121	0.011
13	0.125	0.112	0.011
14	0.013	0.016	0.004
15	0.224	0.241	0.014
16	0.075	0.067	0.008
17	0.075	0.068	0.008
18	0.137	0.138	0.011
19	0.137	0.142	0.012

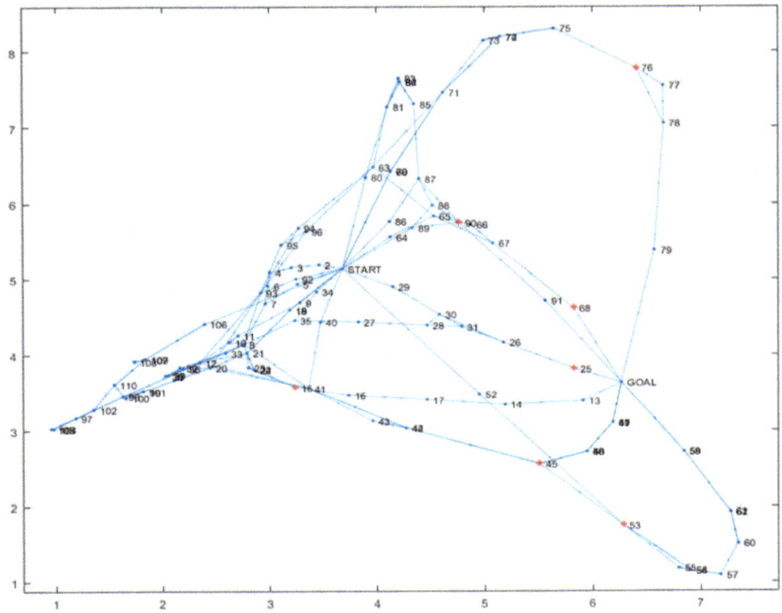

Fig. 5.7 Example 4: a 112-node network

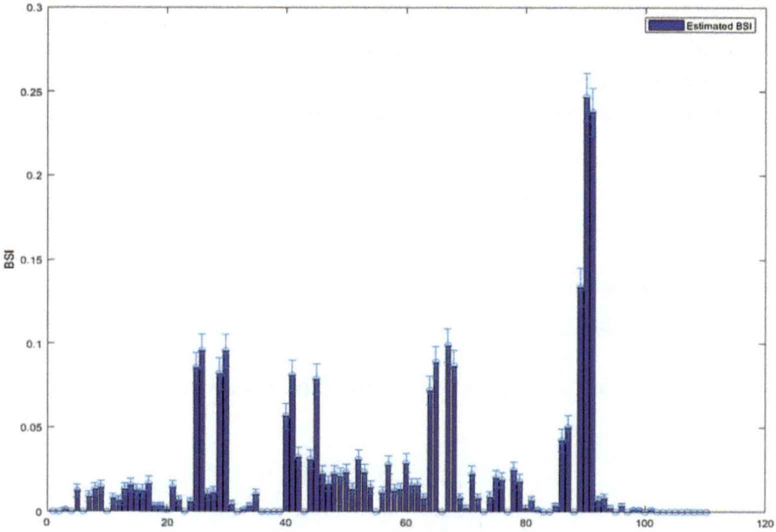

Fig. 5.8 The nodes with the highest BSI scores highlights

5.5 Conclusion

This chapter enhanced the potential use of the Birnbaum Structural Importance measure, a classical measure of nodal criticality, to mission networks consisting of physical systems, which could disrupt an operational pathway of functional components. This was done by (i) introducing a new representation of a mission amenable to BSI analysis consisting of functional nodes and physical nodes that supported the functional nodes and (ii) deriving an approximation with known bounds, ϵ, which were shown theoretically and empirically to be orders of magnitude faster than an exact calculation—$O(2^n)$ vs. $O(\epsilon^{-2}N^2)$. In addition, this chapter introduced a new fast approach for finding the nodes with the top m BSI scores, a new special class of cutsets related to BSI, which was very quick to compute, and a new measure of resilience of a network.

Appendix: BSI as Coefficients of a Linear Regression

While the Birnbaum Structural Importance (BSI) metric for node n, B_n, is classically interpreted as the probability that a given state vector is a critical path vector for node n, assuming all state vectors are equally likely to occur (Birnbaum 1968), there is a simpler linear regression interpretation of B_n.

The linear regression interpretation starts with the challenge of approximating the structural function of the system, $\phi(x)$, with an affine function of $\vec{\beta}$. Recall that the domain of $\phi(x)$, $x = (x_1, x_2, \ldots, x_N)$, is the operability, i.e., the value 0 or 1 of each physical node of the system. The range of $\phi(x)$ is 0 or 1 depending on whether or not the system will be operable given x.

The structural function, $\phi(x)$, can be approximated by

$$\phi(x) \approx \sum_m \beta_m \left(x_m - \frac{1}{2} \right). \tag{5.16}$$

This approximation assumes that $\phi(x)$ has a contribution of $\frac{\beta_m}{2}$ if $x_m = 1$ and $\frac{-\beta_m}{2}$ if $x_m = 0$. As such β_n may be interpreted as the importance of x_n, the state of node n.

Following the classical approach to solving the linear regression of (5.16), one must minimize

$$J(\vec{\beta}) = \sum_x \left(\left[\sum_m \beta_m \left(x_m - \frac{1}{2} \right) \right] - \phi(x) \right)^2, \tag{5.17}$$

where the outer summation is over the 2^N different possible states of $x = (x_1, x_2, \ldots, x_N)$.

The linear regression solution sets $\frac{dJ(\vec{\beta})}{d\beta_n} = 0$, i.e.,

$$0 = \frac{dJ(\vec{\beta})}{d\beta_n} = \sum_x \left(\left[2\left(x_n - \frac{1}{2} \right) \sum_m \beta_m^* \left(x_m - \frac{1}{2} \right) \right] - 2\left(x_n - \frac{1}{2} \right) \phi(x) \right). \tag{5.18}$$

Note that for a fixed value of m and n,

$$\sum_x 2\left(x_n - \frac{1}{2} \right) \beta_m^* \left(x_m - \frac{1}{2} \right) = \begin{cases} 0, & m \neq n, \\ 2^{N-1}\beta_n^*, & n = m. \end{cases} \tag{5.19}$$

This can be seen as follows: The first case of (5.19) follows from the observation that, in the summation, half of the x values have $(x_m, x_n) = (1, 1)$ or $(0, 0)$ making

$$\left(x_m - \frac{1}{2} \right) \left(x_n - \frac{1}{2} \right) = \frac{1}{4}.$$

The other half of the x's in the summation have the property that $(x_m, x_n) = (1, 0)$ or $(0, 1)$ making

$$\left(x_m - \frac{1}{2} \right) \left(x_n - \frac{1}{2} \right) = -\frac{1}{4}.$$

Since the number of positive and negative terms is equal, they cancel in summation.

The second case of (5.19) follows from the observation that each of the 2^N values of x has the property that $(x_n - \frac{1}{2})^2 = \frac{1}{4}$.

Hence, from (5.18) and (5.19),

$$0 = \frac{dJ(\vec{\beta})}{d\beta_n} = 2^{N-1}\beta_n^* - \sum_x \phi(x)(2x_n - 1), \tag{5.20}$$

so

$$\beta_n^* = 2^{-(N-1)} \sum_x \left[\phi(x)(2x_n - 1)\right], \qquad \forall n, \tag{5.21}$$

which, indeed, is the Birnbaum Structural Importance formula for node n, B_n, c.f., (5.5).[3]

References

Ahuja RK, Magnanti TL, Orlin JB (1993) Network flows: theory, algorithms, and applications. Prentice Hall, Upper Saddle River, NJ

Albert R, Albert I, Nakarado GL (2004) Structural vulnerability of the North American power grid. Phys Rev E 69(2):025103

Bienaymé I-J (1853) Considérations à l'appui de la découverte de Laplace. C R Acad Sci 37:309–324

Birnbaum ZW (1968) On the importance of different components in a multicomponent system. Technical Report TR-54, Laboratory of Statistical Research, University of Washington, Washington

Costa LdF, Rodrigues FA, Travieso G, Villas Boas PR (2007) Characterization of complex networks: a survey of measurements. Adv Phys 56(1):167–242

Fussell JB (1975) How to hand-calculate system reliability and safety characteristics. IEEE Trans Reliab 24(3):169–174

Garvey PR, Pinto CA (2009) Introduction to functional dependency network analysis. In: Proceedings of the Second International Symposium on Engineering Systems, Massachusetts Institute of Technology, Cambridge, MA

Garvey PR, Pinto CA, Santos JR (2014) Modelling and measuring the operability of interdependent systems and systems of systems: advances in methods and applications. Int J Syst Syst Eng 5(1):1–24

Guariniello C, DeLaurentis D (2017) Supporting design via the System Operational Dependency Analysis methodology. Res Eng Des 28(1):53–69

Meng FC (2000) Relationships of Fussell–Vesely and Birnbaum importance to structural importance in coherent systems. Reliab Eng Syst Saf 67(1):55–60

Nagurney A, Qiang Q (2008) A network efficiency measure with application to critical infrastructure networks. J Glob Optim 40(1):261–275

Papadimitriou CH, Steiglitz K (1998) Graph search. In: Combinatorial optimization: algorithms and complexity. Dover, Mineola, NY, pp 194–200

[3] For completeness, one must also confirm that $\frac{dJ(\vec{\beta})^2}{d^2\beta_n} = 2(x_n - \frac{1}{2})^2 > 0$.

Rausand M, Høyland A (2004) System reliability theory: models, statistical methods, and applications, 2nd edn. Wiley, New York

Sapir E (1931) Wanted: a world language. Am Mercury 22(86):202–209

Saw JG, Yang MCK, Mo TC (1984) Chebyshev inequality with estimated mean and variance. Am Stat 38(2):130–132

Servi LD, Garvey PR (2017) Deriving global criticality conditions from local dependencies using functional dependency network analysis (FDNA). Syst Eng 20(4):297–306

Shen Y, Nguyen NP, Xuan Y, Thai MT (2013) On the discovery of critical links and nodes for assessing network vulnerability. IEEE/ACM Trans Netw 21(3):963–973

Stoer M, Wagner F (1997) A simple min-cut algorithm. J ACM 44(4):585–591

Tchebychev P (1867) Des valeurs moyennes. J de mathématiques pures et appliquées 12(2):158–176

van der Borst M, Schoonakker H (2001) An overview of PSA importance measures. Reliab Eng Syst Saf 72(3):241–245

Ventresca M, Aleman D (2015) Efficiently identifying critical nodes in large complex networks. Comput Soc Netw 2(1):6

Vesely WE (1970) A time-dependent methodology for fault tree evaluation. Nucl Eng Des 13(2):337–360

Wasserman S, Faust K (1994) Social network analysis: methods and applications. Cambridge University, Cambridge

Chapter 6
Traversing a Network of Physical Nodes to Attack Functional Components of a Mission Dependency Network

A sentence can be a sentence without nouns or adjectives, but never without a verb. For the most part.

Alan Burdick (2018)

6.1 Introduction

In the previous two chapters, we examined the crucial role that both functional and physical components, or nodes, play in the success of a mission network. We modeled physical nodes, typically specific systems, as the enablers, which are necessary to execute functional tasks, where the mission network is represented by a series of functional tasks that must be completed in sequence. In short, the physical nodes are nouns, and the functional nodes are verbs. The degradation of physical nodes ultimately leads to the degradation of the functional nodes they support and consequently the overall effectiveness of the mission network.

This chapter explores the relationship between functional and physical nodes through a different perspective. Here, a physical network must be traversed to gain access to physical nodes that also serve as functional nodes. In order to degrade a mission, this chapter assumes one must identify which physical node of the network to enter, the physical node pathway to traverse, and the physical nodes (which are also functional nodes) to attack. By degrading physical nodes that also serve as functional nodes, one can degrade other dependent functional nodes, ultimately impacting the network's mission. This analysis will be conducted within the context of budget constraints and trade-offs between observing an adversary's traffic and diminishing their mission network.

This chapter is organized as follows. Section 6.2 introduces alternative approaches to characterizing the Measure of Effectiveness (MOE) of the mission network. The specific analysis of interest is introduced by first presenting three simplified variations. For each variation, a description is provided, followed by a simple illustrative example. Section 6.3 presents the detailed analysis of each variation. The chapter ends with Sect. 6.4 summarizing the chapter.

L. Servi, *Mission Dependency Network Analysis*, Risk, Systems and Decisions,
https://doi.org/10.1007/978-3-032-11260-6_6

This chapter introduces three considerations to mission network analysis not addressed in the previous chapters of this book: (1) physical nodes that are part of a physical network that must be traversed in order to impact functional nodes, (2) varying difficulty levels of traversing different edges of the physical network modeled with differing costs, and (3) the possibility of not degrading a physical node even when able to, if the value of observing traffic on the physical node is more desirable.

6.2 Background

6.2.1 The Network MOE and the Propagation of Operability Among the Functional Nodes

The Measure of Effectiveness (MOE) of the mission network and all its functional nodes is a metric between 0 and 100, where 0 corresponds to being inoperable and 100 corresponds to being fully functional. Most functional nodes are impacted by other functional nodes. Those that are not are referred to as the *leaf nodes* of the functional network. These leaf nodes, in this chapter, are typically the ones that can be attacked, as they are also reachable physical nodes.

The decision of which leaf nodes to attack is captured by $\vec{w} \in [0, 1]^{|w|}$, where the *i*th component of $\vec{w}$, $w_i \in \{0, 1\}$, is either 1 corresponding to the decision for leaf node i to be attacked or 0 corresponding to it not attacked (and $|w|$ denotes the number of leaf nodes in the network). If attacked, leaf node i's operability is modified. The ultimate impact of $\vec{w}$ on the MOE is given by $moe(\vec{w}) \in [0, 100]$.

The operability of the functional nodes determines the $moe(\vec{w})$. There are numerous possibilities for the functional form of $moe(\vec{w})$, which relates attack decisions on functional leaf nodes to ultimately impacting the mission network, e.g., Functional Dependency Network Analysis (FDNA) (Garvey and Pinto 2009, 2012; Garvey et al. 2014; Servi and Garvey 2017), the Systems Operational Dependency Analysis (SODA) (Guariniello and DeLaurentis 2017), the near-universal mission dependency framework introduced in Chap. 2, or combinations of AND and OR nodes. For each choice, user-defined parameters model the dependency relationship between child and parent nodes. For illustrative purposes, details of the FDNA equations are given below. However, any dependency representation of a mission network could be used for the analysis in this chapter.

Here, each node is said to have an operability level between 0 and 100, where the operability of the root node, i.e., the node that characterizes the mission, is the MOE, $moe(\vec{w})$. The FDNA equations have two parameters for each arc from a node i to a node j dependent on it: $\alpha_{ij} \in [0, 1]$ and $\beta_{ij} \in [0, 100]$.

If P_i is the operability of functional node i, the recursive FDNA equation is

$$P_i = \begin{cases} \min\left[100 - \frac{1}{|\mathcal{J}(i)|}\sum_{j\in\mathcal{J}(i)}\left(100 - P_j\right)\alpha_{ij}, \min_{j\in\mathcal{J}(i)}\left[P_j + \beta_{ij}\right]\right] & i \notin \mathcal{L} \\ P_i^L & i \in \mathcal{L}, \end{cases} \tag{6.1}$$

where $\mathcal{J}(i)$ is the set of functional nodes that directly impact the operability of node i, i.e., the children of node i, $\mathcal{L}$ is the set of leaf nodes, i.e., the nodes for which $\mathcal{J}(i)$ is the empty set, and P_i^L is the value of leaf node i if not attacked.

One might interpret (6.1) as two distinctive ways that a node's operability is impacted by other nodes. The first, using the α_{ij} parameter, captures the weighted average impact of such nodes. The second, using the β_{ij}, captures the possibility that a single node's operability can directly impact another. Note that the nonlinear equation (6.1) is related to two separate sets of linear inequality constraints: (i) $P_i \leq 100 - \frac{1}{|\mathcal{J}(i)|}\sum_{j\in\mathcal{J}(i)} P_j$ and (ii) $P_i \leq P_j + \beta_{ij} \quad \forall j \in \mathcal{J}(i)$.

One challenge with the analysis of networks is that the required parameters are often not known precisely. Chapters 7 and 8 explore methods to augment the optimization formulation to permit one to find the best decisions that balance the trade-offs of optimizing the objective function and hedging for what is only imperfectly known about the descriptive parameters that capture the network dependencies. This was explicitly explored for FDNA but could be generalized to other $moe(\vec{w})$ functions.

6.2.2 Four-Node Attack Variations

To illustrate the variations and their differences, Fig. 6.1a shows an example of a mission network, and Fig. 6.1b–e present four-node attack variations. For illustrative purposes and without loss of generality, the example assumes there is only one entry node into the physical network. The node at the bottom, node 1, is the sole node from which a physical attack might originate. The rectangular nodes represent purely physical nodes. These are the nodes which must be traversed to impact the network mission. The hexagonal nodes represent physical nodes in the physical network which are also functional nodes, the leaf nodes, of the functional network used to compute $moe(\vec{w})$. These are the nodes which, if attacked, directly impact the network mission. The oval nodes represent the non-leaf functional nodes of a functional network that will transmit the impact of attacks on the leaf nodes to the network mission to compute $moe(\vec{w})$. There also exists a physical node that acts as an observer node, node 2, whose observing ability is denoted by being shaded. This observing physical node can observe traffic originating at node 105, which goes through node 2. The arrows represent either a dependency relationship between functional nodes or traversing options of physical nodes. As will be discussed below for each variation, the optimal attack solution is illustrated with node IDs bolded if a physical node should be disabled and a dashed link if the link should be traversed.

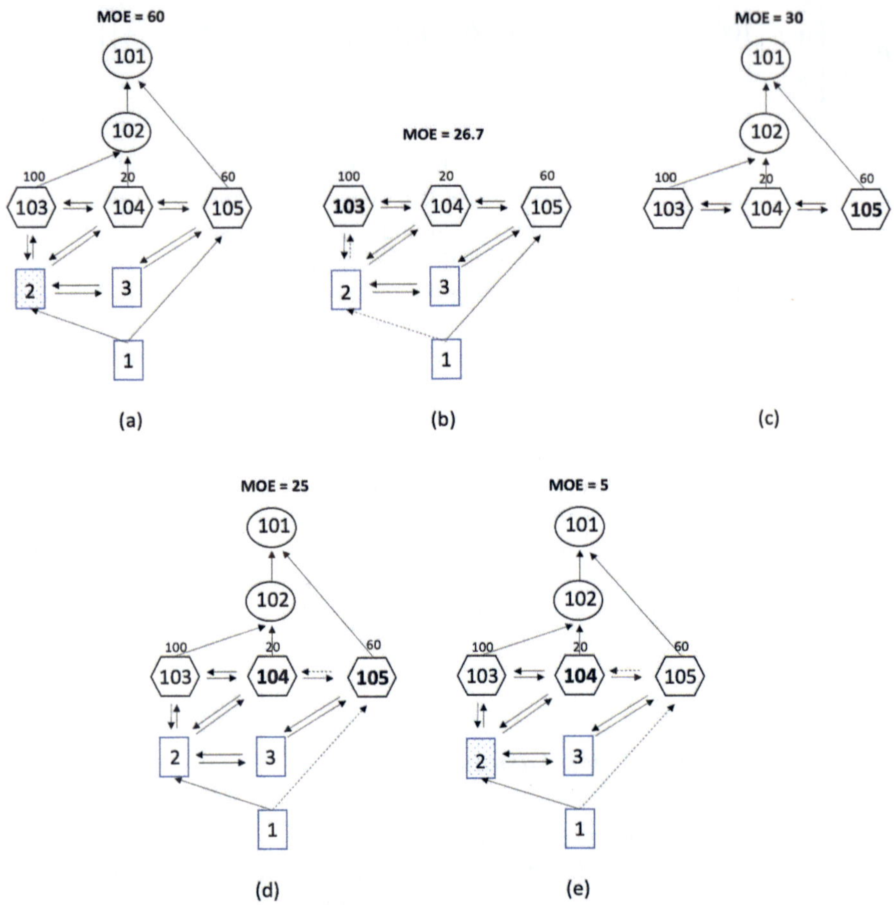

Fig. 6.1 Illustrative examples of node attack variations

For each variation, we assume, without loss of generality, the cost to traverse a link between two physical nodes (one or both of which might also be a functional leaf node) is 1. In practice, this cost is determined by actual network artifacts such as the strength of a firewall and will vary from link to link. Finding weaker points of traversal and taking advantage of already compromised paths can be advantageous to the attacker. The budget for traversing a link is assumed to be 2. The operability of each node is the average of the operability of its dependent children (for FDNA, this corresponds to $\beta = 100$; $\alpha = 1$ for all links). The operability of the leaf nodes if not attacked is 100, 20, and 60 for nodes 103, 104, and 105, respectively, and is displayed in Fig. 6.1. If traffic originating at node 105 passes through the observer, node 2, the traffic is observed, so the MOE should decrease by 50 to reflect this violation; otherwise the MOE is not impacted by the observer node.

The four-node attack variations considered are as follows:

1. **Physical node attack (P1)**, Fig. 6.1b: A network comprised entirely of physical nodes (i.e., it has no non-leaf functional nodes or observer nodes), and the defender has a very simple mission objective: maximize the average of the operability of leaf nodes.
   Description: The attacker starts outside a network, selects a point of entry, which for our example must be node 1, and then spends its budget to traverse the physical arcs in order to reach and destroy (i.e., setting the operability to zero) physical nodes that, for other node attack variations, are also leaf node of a functional network. Since the mission objective is the average operability of the leaf nodes, the value of setting the operability of a leaf node to 0 is the nominal leaf node operability divided by the number of leaf nodes, three. The attacker's goal is then to minimize this value subject to a budget constraint.
   The classical Steiner Prize problem (Hauptmann 2013; Du and Hu 2008; Hajiaghayi et al. 2009) seeks to find a path in a network for which some nodes have "prizes," and the attacker's goal is to maximize the total value of collected "prizes" subject to a budget constraint. Given that P1 uses a mission equal to the simple average of the leaf nodes' operabilities, it is equivalent to the classical Steiner Prize problem where each "prize" has the value of its nominal operability divided by the number of leaf nodes.
   The example: When implementing this physical node attack variation, there is no associated network of functional nodes, so the MOE goal is set to simply minimize the average operability of the leaf nodes. The best attack "collects" as much operability as is possible subject to the budget constraint. For the example in Fig. 6.1b, the budget constrains the attack path to visit no more than two nodes. The optimal solution, shown in Fig. 6.1b, attacks node 103, using the path from node 1 to node 2 and to node 103. Using the simple average function for the MOE, it appears that the MOE is reduced from $(100 + 20 + 60)/3 = 60$ to $(0 + 20 + 60)/3 = 26.7$, and no other attack constrained by a budget could do better.
   However, this would not be an optimal solution if the functional nodes were included. In Fig. 6.1a, the functional nodes would imply that the operability of node 101 is the average of the operability of node 102 and node 105, and the operability of node 102 is the average of the operability of node 103 and node 104. Hence, the solution of Fig. 6.1b is suboptimal, as it overestimates the value of node 103. Specifically, the MOE would not be 26.7, but rather $((0 + 20)/2 + 60)/2 = 35$ (where the first term is the operability of node 102, and the second is the operability of node 105). If instead the attacker chose the path from node 1 to node 105 and then to node 104 and made both node 105 and node 104 inoperable, the MOE would have been $((100+0)/2+0)/2 = 25 < 35$. Hence, this decision would have been better.
2. **Functional node attack (P2)**, Fig. 6.1c: a network with functional nodes but no physical nodes. Its leaf nodes, when attacked, impact the mission objective in specified ways governed by the functional nodes and $moe(\vec{w})$.

Description: Here, the attacker makes decisions to disable the leaf nodes of a functional network, denoted by $\vec{w}$, to minimize the mission MOE, $moe(\vec{w})$, while not exceeding a cost budget.
The example: The functional node attack replaces the physical network traversal costs with the minimal cost for attacking each node in the physical network. Thus, node 103 costs 2 to attack, as it requires two hops from node 1; node 104 costs 2 to attack; and node 105 costs 1 to attack. Since the budget is 2, exactly one leaf node can be attacked. Figure 6.1c shows the optimal solution, which is now to attack node 105. This is because attacking node 105 results in an MOE of $((100+20)/2+0)/2 = 30$, which is smaller than the MOE of attacking node 103, which is $((0+20)/2+60)/2 = 35$, or that of attacking node 104, which is $((100+0)/2+60)/2 = 55$.
However, as noted with the first example, this is not optimal as the marginal cost to attack node 104 and node 105 depends on which other nodes are attacked. Therefore, if costs were properly computed within a budget of 2, it is possible to attack node 105 and node 104, which would result in a lower MOE: $((100+0)/2+0)/2 = 25 < 55$.

3. **Physical and functional node attack (P3)**, Fig. 6.1d: a network with both physical and functional nodes (i.e., a combination of both P1 and P2).
Description: Here, the combination of P1 and P2 replaces the inappropriately simple MOE of P1 and the less precise cost computation of P2. Both prevented the previously found optimal solution. The key to unlocking P3 is to generalize physical node attack formulation to use the more general objective function $moe(\vec{w})$, instead of a summation of prizes collected.
The example: The physical and functional node attack rectifies the mistakes made by the physical node attack and functional node attack: It does not overvalue node 103, and it recognizes that it costs less to attack node 104 after node 105 has been attacked. Figure 6.1d shows the optimal solution to the physical and functional node attack, where after node 105 has been attacked, the attacker continues to attack node 104. The MOE for this attack is $((100+0)/2+0)/2 = 25$.
However, this would not be the optimal solution if one included the impact of the value of node 2 observing traffic originating in node 105 and passing through node 2. Since the proposed solution includes disabling node 105, the attacker cannot gain the benefit of observing traffic originating from node 105.
4. **Physical and functional node attack with observation (P4)**, Fig. 6.1e: a network comprised of both physical and functional nodes as is in P3 and also with an observer node that can observe traffic originating from node 105.
Description: Here, the observer node forces the attacker to perform a trade-off between (a) disabling a physical node to impact the mission through the functional nodes it is connected to and (b) not diminishing the value of monitoring traffic flowing through the physical nodes which, in a different way, may have material advantages. Specifically, there is now a second component of the objective function that represents the diminished value of network traffic being observed in the modified attacked network. The attacker must now decide whether to disable a compromised node or not, whereas in the previous variation

there was never a reason to not disable a compromised node. Not disabling a node may benefit the attacker by routing more network traffic through the observer node.

The example: The solution shown in Fig. 6.1e is the optimal solution, where the impact of the observer node matters: reach 104 via 105 but attack 104 only. If the attacker wants to observe the traffic originating at node 105, it must not disable it. The traffic that originates from node 105 and terminates at node 103 must now all flow through the observer node 2. In the unattacked network, the shortest path is 105, 104, 103, so the traffic is not observed by the observer placed at node 2. The attacker chooses not to disable node 105, even though it has compromised it, because the assumed value of such observation, 50, was assumed to be high. The MOE is now $((100 + 0)/2) + 60)/2 - 50 = 5$, where the first term represents the MOE due to the functionality of the mission, and the second term represents the diminished value of the MOE due to the observed traffic originating at node 105 and passing through the observer at node 2.

The above node attack variations extend the classical use of functional dependency analysis by accounting for more complex considerations. In particular, it incorporates physical realities of traversing through a network to attack functions that impact the mission and the true costs of such an attack. This is in contrast to the more simplistic assumption of each leaf node of the functional network tree having a set cost to attack. This allows attacks to share attacking resources for multiple attacks simultaneously. Finally, this chapter adds further extensions by introducing the trade-offs between disabling a node to impair the mission and allowing a node to not be disabled to permit, and perhaps redirect traffic, which can be observed.

6.3 Analysis

The mission network is denoted as $D = (N, A)$, a directed network, where N is the set of physical nodes, and A is the set of the physical links that an attack can occur over. The node set $N_0 \subseteq N$ represents the set of potential entry nodes to the physical network for any attacks. The proposed definition of an attack is that an *attack path* to node $n_k \in N$ is the sequence of nodes $n_0, n_1, n_2, \ldots, n_{k-1}, n_k$, where $(n_{i-1}, n_i) \in A$. The nodes $n_k \notin N_0$ for $k \neq 0$ and $n_0 \in N_0$ are said to be *compromised* if it is on any attack path.

There are some assumptions regarding how attack paths are made. The first assumption is that attacks are not cumulative, i.e., a node attacked twice is compromised only once. Therefore, each leaf node needs only to be attacked by a single attack path. The second assumption is that once a node is compromised; an attack can be made on any incident node. Any compromised node may be *disabled* freely by the attacker. In variations **(P1)**, **(P2)** and **(P3)**, it is assumed that all compromised nodes are disabled nodes. In variation **(P4)**, it can be seen that it might be desirable to compromise a node, but not disable it, as a means of routing network traffic advantageously.

Table 6.1 Set notation

Set	Condition	Description
N		Set of physical nodes
N_0	$N_0 \subseteq N$	Node set representing external attack entries to the network
$\mathcal{L}$	$\mathcal{L} \subseteq N, N_0 \cap \mathcal{L} = \emptyset$	Leaf nodes
A		Set of arcs attacks can occur over
N_M	$N_M \subseteq N$	Nodes containing observers of valued traffic
A_M	$A_M \subseteq (N \backslash N_0) \times (N \backslash N_0)$	Set of node pairs (s, t) with valued traffic

Table 6.2 Parameter notation

Parameter	Condition	Description
c_a	$a \in A$	Cost to attack via arc a
v_n	$n \in \mathcal{L}$	Value of disabling node n
b		Budget for attacks

Table 6.3 Variable notation

Variable	Condition	Description
w_i	$i \in \mathcal{L}, w_i \in \{0, 1\}$	Indicator of attack of a leaf node i
p_a	$a \in A, p_a \in \{0, 1\}$	Indicator of attack of arc a
$u_{k,a}$	$k \in \mathcal{L}, a \in A, u_{k,a} \in \{0, 1\}$	Attack path indicator for each leaf node
$g_a^{s,t}$ or g_a	$(s, t) \in A_M, a \in A, g_{(s,t),a} \geq 0$	Shortest path flow
$x_i^{s,t}$ or x_i	$(s, t) \in A_M, i \in N \backslash N_0$	Shortest path dual values

Tables 6.1, 6.2 and 6.3 defines the notation shared by the four node attack variations.

6.3.1 *Analysis of Physical Node Attack (P1)*

As introduced in Sect. 6.2, the physical node attack variation consists of a network of physical nodes, but no exclusively functional nodes or observer nodes. The goal is to identify the best coordinated attack paths from outside the network that may share attack costs by following overlapping paths to separate physical nodes, which are also functional leaf nodes. With no functional dependency network, minimizing the average operability of each leaf node is the assumed objective. This is equivalent to maximizing the sum of operabilities disabled in a Prize Collecting Steiner Tree (PCST) Ljubić et al. (2006). The variable $\mathbf{v}$ is used for this purpose as a prize.

The modeling approach for this variation is that of a multi-commodity single path formulation, although alternative approaches are considered: The Big M flow could have been used to simulate the opening of arcs, but the leaf nodes are assumed to be somewhat sparse, and Big M is notoriously difficult. Subtour-type constraints could have been applied, but in prize collecting scenarios, a high proportion of these constraints can be expected to be active at optimality, so they are unreasonable. A path formulation could have been used and might be strong, but it is a longer implementation process.

In the multi-commodity single path formulation, for every node that is attacked, a path must be identified using a single unit of flow from the set N_0 to the node. If leaf nodes are relatively sparse or the network is relatively small, this is a particularly reasonable formulation.

$$\min_{p,u,w} \left[\sum_k \frac{v_k(1-w_k)}{|\mathcal{L}|} \right] \tag{6.2}$$

subject to

$$\vec{c}^T \vec{p} \le b, \tag{6.3}$$

$$\sum_{(n_0,j)\in A: n_0 \in N_0} u_{k,(n_0,j)} - w_k = 0, \qquad \forall k \in \mathcal{L}, \tag{6.4}$$

$$\sum_{(i,k)\in A} u_{k,(i,k)} - w_k = 0, \qquad \forall k \in \mathcal{L}, \tag{6.5}$$

$$\sum_{(j,i)\in A} u_{k,(j,i)} - \sum_{(i,j')\in A} u_{k,(i,j')} = 0, \qquad \forall k \in \mathcal{L}, \forall i \in N : i \notin N_0, i \ne k, \tag{6.6}$$

$$u_{k,a} - p_a \le 0, \qquad \forall k \in \mathcal{L}, \forall a \in A, \tag{6.7}$$

$$\vec{p} \in \{0,1\}^{|A|}, \tag{6.8}$$

$$\vec{u} \in \{0,1\}^{|\mathcal{L}|\cdot|A|}, \tag{6.9}$$

$$\vec{w} \in \{0,1\}^{|\mathcal{L}|}. \tag{6.10}$$

The model will find one or more attack paths that start outside the network, enter via a physical entry node to the network, $n_0 \in N_0$, and end at a physical node k that is also a functional leaf node that is disabled. For each path that is chosen, all of its arcs must be used for an attack. Note that optimality guarantees a node would never need to be attacked more than once.

More specifically, one can interpret (6.2)–(6.10) as follows: Since $u_{k,\alpha}$ corresponds to whether or not link α is part of the attack path to node k, (6.3) and (6.7) ensure that the cost is incurred for each arc used and, implicitly, that unimpactful arcs are not used. Equations (6.4) and (6.5) insist that each leaf node k attacked will start from a physical entry node, i.e., a node in the set N_0 and ends with an arc to node k. Equation (6.6) states that each node on the path to leaf node k has an outgoing arc if it has an incoming arc. Finally, (6.2) states that the objective of the attacker is to minimize the average of the operability of the functional leaf nodes not disabled. This is equivalent to maximizing the "prizes," denoted as v_k, that can be gathered within the budget constraint, in (6.3).

6.3.2 Functional Node Attack (P2)

For the physical node attack variation (P1), each physical arc has a known cost associated with it, so the cost of disabling a functional leaf node was the summation of the cost of the path to that leaf node along the physical arcs. Similarly, for the functional node attack variation (P2), each leaf node to a functional decision tree has a cost to attack and disable. However, for the functional node attack variation, there are no physical arcs, so the cost of disabling a leaf node must be provided. This *nodal* cost will be denoted by $\bar{\mathbf{c}}$ to differentiate it from the *arc* costs, $\mathbf{c}$, for the functional node attack variation.

Here, the MOE is minimized subject to the constraint that the sum of the nodal costs must be no more than the budget. The provided nodal costs, $\bar{\mathbf{c}}$, for node i might, for example, be assigned to be cost of least costly pathway from outside the network to node i using the arc costs $\mathbf{c}$. This attack variation can be formulated as

$$\min_{w} \left[moe(\vec{w}) \right] \tag{6.11}$$

subject to

$$\vec{\bar{c}}^{\,T} \vec{w} \leq b, \tag{6.12}$$

$$\vec{w} \in \{0, 1\}^{|\mathcal{L}|}, \tag{6.13}$$

where the value of the mission is determined by the MOE function, moe($\vec{w}$), based on the leaf nodes attacked. This may be from a combination of simple averages at each node, from FDNA (Garvey and Pinto 2009, 2012; Garvey et al. 2014; Servi and Garvey 2017), or could be derived from any function. Section 6.2.1 describes in more detail the analysis if FDNA is used. Chapters 7 and 8 describe how to modify the analysis for the case in which FDNA parameters are known imperfectly.

6.3.3 Physical and Functional Node Attack (P3)

The physical and functional node attack variation combines the physical nodes of (P1) with the functional nodes of (P2). This problem is closely related to PCST problems, except that rather than a simple prize, the objective is based on the $moe(\vec{w})$, which captures the aggregation of the functional nodes. This can be formulated as

$$\min_{p,u,w} \left[moe(\vec{w}) \right] \tag{6.14}$$

subject to

$$\vec{c}^T \vec{p} \leq b, \tag{6.15}$$

$$\sum_{(n_0,j)\in A: n_0 \in N_0} u_{k,(n_0,j)} - w_k = 0, \qquad \forall k \in \mathcal{L}, \tag{6.16}$$

$$\sum_{(i,k)\in A} u_{k,(i,k)} - w_k = 0, \qquad \forall k \in \mathcal{L}, \tag{6.17}$$

$$\sum_{(j,i)\in A} u_{k,(j,i)} - \sum_{(i,j')\in A} u_{k,(i,j')} = 0, \qquad \forall k \in \mathcal{L}, \forall i \in N : i \notin N_0, i \neq k, \tag{6.18}$$

$$u_{k,a} - p_a \leq 0, \qquad \forall k \in \mathcal{L}, \forall a \in A, \tag{6.19}$$

$$\vec{p} \in \{0, 1\}^{|A|}, \tag{6.20}$$

$$\vec{u} \in \{0, 1\}^{|\mathcal{L}|\cdot|A|}, \tag{6.21}$$

$$\vec{w} \in \{0, 1\}^{|\mathcal{L}|}. \tag{6.22}$$

The interpretation of (6.14)–(6.22) is identical to that of (6.2)–(6.10), except now the objective function is different. In many cases, for example, when $moe(\vec{w})$ is an FDNA evaluation, the optimization can be put in a linear form (see Sect. 6.2.1).

6.3.4 Physical and Functional Node Attack with Observation (P4)

In addition to both physical and functional nodes, as is in P3, this variation also consists of one or more observer nodes that have the ability to observe traffic originating from physical nodes and passing through the observer nodes. The analysis is more complicated than (P1), (P2), and (P3). This variation is a combination of the others. It acknowledges that an attacked node may have the ability to observe network traffic. Hence, destroying a node may have the collateral damage of limiting their ability to gather potentially valuable information. The analysis of this variation must expand the physical and functional node attack problem (P3) with a second objective, which acknowledges the value of the ability to observe network traffic from nodes in $N \backslash N_0$, which have *observers* in them. To simplify the analysis in a conservative way and with limited informational requirements, it will be assumed that (i) network traffic initiating and terminating in N does not travel outside N, (ii) all arcs are uncapacitated, and (iii) network traffic between two nodes will always take the resulting shortest path using nodes that have not been disabled but, in case of a tie, will not pass through a node that is also an observer.

The sets, parameters, and variables required for the analysis are listed in Sect. 6.3. Of special note, N_M is the set of nodes that can observe traffic, and A_M is the set of node source/termination pairs, (s, t), defining a logical connection for which observing the traffic is of value. It is assumed that a node in N_M will observe valued traffic between nodes s and t if (1) $(s, t) \in A_M$ and (2) the shortest path between nodes s and t along nondisabled nodes goes through the node in N_M.

The primal formulation of the shortest path from s to t in terms of the number of hops across nodes that have not been disabled is given by

$$\min_{g} \left[\vec{1}^T \vec{g} \right] \tag{6.23}$$

subject to

$$\sum_{(j',k)\in A, w_{j'}=0, j'\notin N_0} g_{j',k} - \sum_{(k,j'')\in A, w_{j''}=0, j''\notin N_0} g_{k,j''} = h_k,$$

$$\forall k \in \{k : k \in N\backslash N_0, w_k = 0\}, \tag{6.24}$$

$$\vec{g} \geq \vec{0}, \tag{6.25}$$

where

$$h_k = \begin{cases} -1, & k = s, \\ 1, & k = t, \\ 0, & \text{else.} \end{cases} \tag{6.26}$$

This implicitly assumes, without loss of generality, that all communications between sources and destinations have a flow of 1. Equation (6.23) states that the number of hops is being minimized. Equation (6.24) states that node s has one more outgoing arc than incoming arcs, t has one more incoming arc than outgoing arcs, and the rest of the nodes that have not been disabled have equal numbers of incoming and outgoing arcs.

Since $g_{i,j}$ appears in the left part of (6.24) for $k = j$ and the right part of (6.24) for $k = i$, the dual problem is

$$\max_{x} \left[x_t - x_s \right] \tag{6.27}$$

subject to

$$x_j - x_i \leq 1, \qquad \forall (i, j) \in \{(i, j) \in A, w_i = w_j = 0, \ \forall i \notin N_0, \forall j \notin N_0\}. \tag{6.28}$$

Equation (6.24) is equivalent to (6.29)–(6.31), i.e., the summation in (6.24) can be simplified to (6.29) if the constraints (6.30) and (6.31) are added.

$$\sum_{(j',k)\in A, j'\notin N_0} g_{j',k} - \sum_{(k,j'')\in A, j''\notin N_0} g_{k,j''} = h_k, \qquad \forall k \in \{k : k \in N \backslash N_0\}, \tag{6.29}$$

$$g_{j',k} + w_{j'} \leq 1, \qquad \forall j', k \in N \backslash N_0 : (j', k) \in A, \tag{6.30}$$

$$g_{k,j''} + w_{j''} \leq 1, \qquad \forall j'', k \in N \backslash N_0 : (k, j'') \in A. \tag{6.31}$$

This follows because (6.30) forces $g_{j',k}$ to 0 if $w_{j'}$ is 1, i.e., node k is disabled. Similarly, (6.31) forces $g_{k,j''}$ to 0 if $w_{j''} = 1$.

Correspondingly, (6.28) is equivalent to

$$x_j - x_i \leq 1, \qquad \forall (i, j) \in \{(i, j) \in A, i \notin N_0, j \notin N_0\}, \tag{6.32}$$

$$x_j - x_i - M(w_i + w_j) \leq 1, \qquad \forall (i, j) \in \{(i, j) \in A, i \notin N_0, j \notin N_0\}, \tag{6.33}$$

where M is a big number, e.g., $|N \backslash N_0|$. This follows because (6.33) effectively removes any constraint on $x_j - x_i$ if either w_i or w_j is 1. This would be the case if either node i or node j were disabled.

The above analysis does not consider the possibility that the shortest path is infeasible, either due to s or t being attacked or due to s and t having a cutset being attacked, i.e., (6.29) might be infeasible. One approach to handling this is to modify the role of h_k by accounting for whether the traffic from s to t going through k is being observed. Specifically, notation $g_{i,j}$ is modified to be $g_{i,j}^{s,t}$, which will correspond to traffic along the (i, j) arc that originates at s and terminates at t and modify the dual variable x_j to be $x_j^{s,t}$. The premise is that the shortest path will only be calculated if $m_{(s,t)} = 1$, that is, the network traffic passes through an observer. If the $s - t$ pair is cut or the shortest path is not incident with an observer, $m_{(s,t)}$ will be forced to 0 and make the primal/dual trivial with 0 solutions. These modifications can be made by setting

$$h_s = -m_{(s,t)},$$

$$h_t = m_{(s,t)}$$

and then requiring that some observer be incident to the path with

$$\sum_{(i,j)\in A:i,j\in N\setminus N_0, i\in N_M \text{ or } j\in N_M} g_{i,j}^{s,t} - m_{(s,t)} \geq 0,$$

as well as requiring primal/dual equality with

$$\vec{1}^T\vec{g} - (x_t - x_s) = 0.$$

These shortest path models must be done for each $(s,t) \in A_M$. These changes additionally trim out some unnecessary variables, which are always 0 due to optimality conditions. The assumption made on the model that any monitored path is lexicographically longer than an unmonitored path of the same length is not explicitly modeled; perturbations on the primal/dual of the shortest path sub-models can easily achieve this.

The following proposed model combines the physical and functional node attack problem (P3) with a new objective and the discussed shortest path sub-models for each $(s,t) \in A_M$.

$$\min_{p,u,m,g} \left[moe(\vec{m}) - moe_{obs}(\vec{m})) \right] \tag{6.34}$$

subject to

$$\vec{c}^T\vec{p} \leq b, \tag{6.35}$$

$$\sum_{(n_0,j)\in A:n_0\in N_0} u_{k,(n_0,j)} - w_k = 0, \qquad \forall k \in N\setminus N_0, \tag{6.36}$$

$$\sum_{(i,k)\in A} u_{k,(i,k)} - w_k = 0, \qquad \forall k \in N\setminus N_0, \tag{6.37}$$

$$\sum_{(j,i)\in A} u_{k,(j,i)} - \sum_{(i,j')\in A} u_{k,(i,j')} = 0, \qquad \forall k \in N\setminus N_0, \forall i \in N : i \notin N_0, i \neq k, \tag{6.38}$$

$$u_{k,a} - p_a \leq 0, \qquad \forall k \in N\setminus N_0, \forall a \in A, \tag{6.39}$$

$$\sum_{(s,j)\in A, j\notin N_0, s\in N_M \text{ or } j\in N_M} g_{s,j}^{s,t} - m^{s,t} = 0, \qquad \forall (s,t) \in A_M, \tag{6.40}$$

$$\sum_{(i,t)\in A, i\notin N_0, i\in N_M \text{ or } t\in N_M} g_{i,t}^{s,t} - m^{s,t} = 0, \qquad \forall (s,t) \in A_M, \tag{6.41}$$

$$\sum_{(j,i)\in A, j\notin N_0} g_{j,i}^{s,t} - \sum_{(i,j')\in A, j'\notin N_0} g_{i,j'}^{s,t} = 0, \qquad \forall (s,t) \in A_M, \forall i \in N \backslash \{N_0 \cup \{s,t\}\}, \tag{6.42}$$

$$\sum_{(i,j)\in A: i,j\in N\backslash N_0, i\in N_M \text{ or } j\in N_M} g_{i,j}^{s,t} - m^{s,t} \geq 0, \qquad \forall (s,t) \in A_M, \tag{6.43}$$

$$g_{i,j}^{s,t} + w_i \leq 1, \qquad \forall (s,t) \in A_M, \forall i,j \in N \backslash N_0 : (i,j) \in A, \tag{6.44}$$

$$g_{i,j}^{s,t} + w_j \leq 1, \qquad \forall (s,t) \in A_M, \forall i,j \in N \backslash N_0 : (i,j) \in A, \tag{6.45}$$

$$x_j^{s,t} - x_i^{s,t} - |N \backslash N_0|(v_i + v_j) \leq 1, \qquad \forall (s,t) \in A_M, \forall i,j \in N \backslash N_0 : (i,j) \in A, \tag{6.46}$$

$$\mathbf{1}^T \mathbf{g^{s,t}} - (x_t^{s,t} - x_s^{s,t}) = 0, \qquad \forall (s,t) \in A_M, \tag{6.47}$$

$$\vec{p} \in \{0,1\}^{|A|}, \tag{6.48}$$

$$\vec{u} \in \{0,1\}^{|N\backslash N_0|\cdot|A|}, \tag{6.49}$$

$$\vec{m} \in \{0,1\}^{|A_M|}, \tag{6.50}$$

$$\vec{g}^{s,t} \geq \vec{0}, \qquad \forall (s,t). \tag{6.51}$$

The objective function now has an additional component corresponding to the value of the observed traffic. Constraints (6.35)–(6.39) are from the physical node attack variation. (P1) Constraints (6.40), (6.41), and (6.42) are similar path constraints. They require traffic that is observed to have a flow of one from the sender to the receiver as we are assuming, without loss of generalization, that all communication consists of one unit of flow between pairs of nodes. Constraint (6.43) requires the observed traffic be incident with some observer. Constraints (6.44) and (6.45) prohibit network traffic from using a disabled node. Constraints (6.46) and (6.47) form the duals for the shortest paths, enforcing that they are simple, shortest paths. Constraint (6.47) has a Big M coefficient to trivialize constraints that should not be in the dual due to the nodes being disabled.

6.4 Conclusion

In this chapter, an alternative approach to analyzing physical and functional mission networks was introduced, distinct from the approach presented in the previous two chapters. This chapter presented an approach for identifying the optimal entry point into the physical network and the most effective pathway through physical nodes to mission-impactful functional nodes. This required balancing the intelligence gain/loss trade-off between mission impact and the value of information gained by observing network traffic. To build toward this full formulation, (P4), the analysis began with three progressively complex subproblems: (P1) addressed mission networks involving only physical nodes, (P2) focused solely on functional nodes, and (P3) considered both physical and functional nodes without traffic observation capability. These subproblems served as stepping stones that informed the development of the full problem solution.

References

Burdick A (2018) Why nouns slow us down, and why linguistics might be in a bubble. The New Yorker, New York (2018)

Du DZ, Hu X (2008) Steiner tree problems in computer communication networks. World Scientific, New York

Garvey PR, Pinto CA (2009) Introduction to functional dependency network analysis. In: Proceedings of the second international symposium on engineering systems. Massachusetts Institute of Technology, Cambridge, MA. http://esd.mit.edu/symp09/day3.html

Garvey PR, Pinto CA (2012) Advanced risk analysis in engineering enterprise systems. Chapman & Hall/CRC, Boca Raton. ISBN 9781439826140

Garvey PR, Pinto CA, Santos JR (2014) Modelling and measuring the operability of interdependent systems and systems of systems: advances in methods and applications. Int J Syst Syst Eng 5(1):1–24

Guariniello C, DeLaurentis D (2017) Supporting design via the System Operational Dependency Analysis methodology. Res Eng Des 28(1):53–69

Hajiaghayi MT, Kortsarz G, Salavatipour MR (2009) Approximating buy-at-bulk and shallow-light k-Steiner trees. Algorithmica 53(1):89–103

Hauptmann M (2013) A compendium on Steiner tree problems. Research Report 85343-CS, Department of Computer Science, University of Bonn, Bonn, pp 1–44

Ljubić I, Weiskircher R, Pferschy U, Klau GW, Mutzel P, Fischetti M (2006) An algorithmic framework for the exact solution of the prize-collecting Steiner tree problem. Math Program 105(2):427–449

Servi LD, Garvey PR (2017) Deriving global criticality conditions from local dependencies using functional dependency network analysis (FDNA). Syst Eng 20(4):297–306

Part III
Mission Dependency Network Analysis with Incomplete System-of-System Knowledge and Temporal Considerations

Chapter 7
Robust Mission Network Analysis: The Theory

You must never confuse faith that you will prevail in the end—which you can never afford to lose—with the discipline to confront the most brutal facts of your current reality, whatever they might be.

Admiral James Stockdale (Collins 2001)

7.1 Introduction

7.1.1 Overview

Impactful analysis of a mission network defined by its dependency mapping depends heavily on parameters to accurately capture its dependencies. Experience tells us that estimating such parameters precisely is fraught with challenges. Ignoring such realities risks inappropriate recommendations. Chapters 7 and 8 confront this brutal fact by extending existing robust optimization methods to create the capability of defensive mission network analysis which is robust to imprecise input parameter estimates. An analyst should be faulted if they recommend decisions that are not robust to imprecise assumptions. While such parameters are difficult to estimate precisely, there are methods to hedge against such uncertainty.

Optimistic neglect typically has minimal consequences. However, when it has consequences, they can be very large. It might therefore be tempting to be excessively pessimistic and conservative, but that approach suffers from the cost of superfluous hedging. Success is found by balancing these considerations, and this chapter provides an analytical framework to achieve such balance.

Specifically, to prepare the groundwork for analysis of imprecise dependency parameters, this chapter first examines classical linear program optimization problems under the assumption that the parameters defining the problems might not be known precisely but instead only known to be in a specified interval. This chapter gives a brief overview of current approaches to address this lack of knowledge with a focus on the work of Bertsimas and Sim (2004). The theoretical analysis of Bertsimas and Sim (2004) is then extended to be applicable for the challenges

L. Servi, *Mission Dependency Network Analysis*, Risk, Systems and Decisions,
https://doi.org/10.1007/978-3-032-11260-6_7

of most interest in this book. The analysis is then specialized to determine how to allocate constrained resources to defend a mission network by fortifying some nodes while hedging against imperfect knowledge of the parameters which describe the mission network. This uses a single parameter Γ_{tot}, which could be interpreted as how much the decision-maker is willing to trade off mission network protection versus hedging against known unknowns. The approach is illustrated with a small example to illustrate the concepts. The new approach used in this chapter is estimated to reduce the computational requirements by 80 orders of magnitude compared to using Bertsimas and Sim (2004) for small missions with only 50 dependencies.

7.1.2 Technical Overview

This chapter examines the robustness of defensive mission decisions within the context of imperfect information with a Functional Dependency Network Analysis (FDNA) dependency representation. FDNA is chosen for illustrative purposes and was first introduced by Garvey and Pinto (2009) and in Chap. 1. Although this chapter focuses on FDNA, the methods presented in this chapter can be applied to many other directed acyclic dependency representations. While this chapter focuses on defensive mission network analysis, a similar treatment is possible for offensive mission network analysis.

There are multiple approaches to modeling the defense of a mission network. For illustrative purposes, this chapter assumes a user makes defensive mission network analysis decisions to maximize the network's Measure of Effectiveness (MOE) by bolstering individual leaf nodes against potential adversaries. Specifically, this chapter models defenders with resources to make a chosen set of leaf nodes resistant to attack. All other leaf nodes are assumed to be vulnerable to attack. Successfully attacked nodes diminish their MOE values, which propagate throughout the network and ultimately reduce the MOE of the entire network. This chapter describes how the defender can most efficiently deploy such defense resources.

Before examining robust defensive mission network analysis, this chapter detours into an examination of the robust optimization of a generic linear program with coefficients that are only known imperfectly. One approach to robust optimization is to use stochastic programming to optimize the uncertain mission networks such that the solution is feasible with some probability as discussed in Birge and Louveaux (2011) and Prékopa (2013). Stochastic programming requires probability distributions for the uncertainty and that can be more information than the user has. Additionally, this approach allows constraints to be violated. The work (Ben-Tal 1985) attempts to offset this with entropic penalties for solutions that are not "feasible in the mean." The work (Mulvey et al. 1995) uses scenario optimization to find solutions that are *solution robust* (nearly optimal) and *model robust* (nearly feasible). These adjustments still allow the program to be infeasible with large consequences. In this chapter, we examine robust deterministic optimization.

Soyster, in (1973), developed the foundation for needed generalization. He suggested that the optimal solution should remain feasible for all uncertainty in a convex set. Soyster's scenario can be understood from the following linear program:

$$\max_x c^T x \tag{7.1}$$

subject to

$$\sum_j A_j x_j \leq b, \qquad \forall\{A_j \in K_j, \forall j\}, \tag{7.2}$$

$$x \geq 0, \tag{7.3}$$

where c, x, and b are vectors, c^T is the transpose of c, A is a matrix whose jth row is A_j, and the uncertainty sets K_j are convex. This is a natural generalization of the classical linear programming problem (7.4)–(7.6) of determining the value of the vector x that maximizes a linear objective function while subject to nonnegativity constraints and a set of linear constraints. The complication with this linear program is that the coefficients of the linear constraints are not known but instead are only known to be in a known convex set. This columnwise uncertainty problem might be addressed by setting the coefficients equal to the maximum value in the convex set. This ensures that any column vector in the uncertainty set can be used, and the linear program will still be feasible.

Unfortunately, solutions that are feasible for every possible uncertainty are generally too conservative. They may result in a dramatic reduction of optimality from the nominal problem. Therefore, the optimality reduction from the nominal problem due to the inclusion of robustness is unduly excessive. There must be a balance between robustness and optimality. The work (Ben-Tal and Nemirovski 199) addresses this issue by introducing solutions to linear programs with ellipsoidal uncertainties. Although this development reduces the overconservatism inherent in the approach presented in Soyster (1973), ellipsoidal uncertainty is not a form that can be readily applied to mission networks.

The paper (Bertsimas and Sim 2004) made a big leap toward resolving this issue by introducing a parameter for the level of conservatism for each constraint in a linear program. The conservatism parameter Γ_i is the number of coefficients in constraint i that are allowed to vary within a specified interval, cf. (7.14)–(7.20). This viewpoint produces a solution that is feasible in the worst-case scenario for that level of conservatism and is feasible with a high probability if the parameter variation exceeds the chosen level of conservatism. The introduction of Γ_i creates a tuning parameter for the balance between conservatism and robustness. The paper demonstrated that this was achieved with a linear program (7.23)–(7.28).

Using Γ_i, the formulation in Bertsimas and Sim (2004) permits a decision-maker to specify the trade-off between the robustness with respect to coefficient values and their level of conservatism. That this must be specified for *each* constraint makes the level of granularity of this analysis quite fine. Unfortunately, this is too fine a level of granularity for the mission network analysis considered in this book. As a practical matter, the optimization formulation of mission network analysis using FDNA contains many constraints with only one coefficient. Therefore, the formulation in Bertsimas and Sim (2004) would force the decision-maker to choose whether a single coefficient may deviate from its nominal value or not. Instead, we wish to enable the decision-maker to specify a level of conservatism across the *entire* optimization formulation by specifying the total number of coefficients that may deviate from their nominal values across all constraints. For this capability, we must extend the analysis of Bertsimas and Sim (2004) to lessen the granularity and permit the introduction of a parameter for a generalized level of conservatism, Γ_{tot}. The parameter Γ_{tot} is defined as the total number of coefficients in the linear program that may deviate from their nominal value across all constraints.

Section 7.2 describes three steps to extend the approach in Bertsimas and Sim (2004) by introducing a single parameter for the general level of conservatism for the system. Step 1, in Sect. 7.2.1, follows the work of Bertsimas and Sim (2004) to derive robustness with respect to imposing an upper limit on the number of coefficients not equal to their nominal value for *each* individual constraint for a classical linear program optimization. Step 2, in Sect. 7.2.2, removes the restriction of a constraintwise limit, but that removal unfortunately introduces nonlinearity into the formulation, which would impair solution computation speed. Finally, to overcome this impairment, Step 3, in Sect. 7.2.3, derives a new equivalent reformulation that removes the nonlinearity.

Section 7.3.1 demonstrates how this new method is used with a small, illustrative example. Section 7.3.2 describes how a decision-maker might wish to determine the best level of conservatism, Γ_{tot}. In the process of doing so, Sect. 7.3.2 also provides evidence that the resilience of a system could be characterized by a single parameter.

Section 7.4 brings all the analysis together. Sections 7.4.1–7.4.3 frame the nominal network defense analysis problem using FDNA in the absence of any uncertainty of parameters as a classical linear program. Sections 7.4.4 and 7.4.5, and the Appendix apply the three steps of Sects. 7.2.1, 7.2.2, and 7.2.3, respectively, to transform the network defense analysis problem from a nominal one with no uncertainty to one with uncertainty and a parameter to the level of conservatism, Γ_{tot}, to create a robust solution. Finally, Sect. 7.4.7 illustrates the approach with an example.

7.2 Adding Robustness

Robustness is added to a classical linear program (7.4)–(7.6) by introducing the possibility that a specified number of coefficients of A in (7.5) deviate from their assumed nominal values. This is addressed in three steps:

- **Step 1**: *constraining the number of coefficients not equal to their nominal value for each constraint*: Introduce the possibility that the number of deviations is specified for each constraint. This is formally stated in (7.14)–(7.20) and, following an analysis along the lines of Bertsimas and Sim (2004), is shown to be equivalent to (7.23)–(7.28).

- **Step 2**: *constraining the total number of coefficients not equal to their nominal values*: Modify the formulation of Step 1 to assume that the number of deviations needs only to be specified for the entire set of constraints rather than for each constraint individually. This is formulated as (7.29)–(7.35). The key step to the analysis of this new formulation entails finding the dual of (7.23)–(7.28) while now treating Γ_i as a variable, rather than a specified constant, and then adding the constraint $\sum_i \Gamma_i \leq \Gamma_{tot}$. The dual is shown to be (7.38)–(7.43).
- **Step 3**: *linearizing the new formulation*: The natural formulation of Step 2 has the disadvantage that it introduces nonlinearity, the term $\gamma_i \lambda_i$, which will slow down the optimization computation. Therefore, the final step in the analysis is to linearize these equations and add the constraint $\sum_i \Gamma_i \leq \Gamma_{tot}$. This results in the final formulation (7.46)–(7.57).

7.2.1 Step 1: Constraining the Number of Coefficients not Equal to Their Nominal Value for Each Constraint

A classical linear program is defined by the coefficient matrix $A \in \mathcal{R}^{m \times n}$, the variable vector $x \in \mathcal{R}^n$, the vector of the right-hand side constants $b \in \mathcal{R}^m$, and the vector of objective value coefficients $c \in \mathcal{R}^n$.

$$\max_x c^T x \tag{7.4}$$

subject to

$$\sum_j A_j x_j \leq b, \tag{7.5}$$

$$x \geq 0. \tag{7.6}$$

As a reminder, the dual of (7.4)–(7.6) is

$$\min_\phi \; b^T \phi \tag{7.7}$$

subject to

$$A^T \phi \geq c, \tag{7.8}$$

$$\phi \geq 0. \tag{7.9}$$

Duals will be used later in this chapter in (7.22), (7.38)–(7.43), and (7.87)–(7.103).

For robust analysis, we assume there is uncertainty in the A matrix, represented by the matrix Δ, such that the true values $A^* \in \left[\bar{A} - \Delta, \bar{A} + \Delta\right]$. One might view the matrix $\bar{A}$, whose components are $\bar{a}_{ij}$, as the nominal, best estimate, or baseline value for A and the matrix δ, componentwise, as the width of uncertainty of the value of A^*.

It might appear restrictive to not assume potential uncertainty of b or c and only focus on the uncertainty of the matrix A. As noted by Bertsimas and Sim (2004), this is not a restriction, as one can always modify the problem formulation to put b or c uncertainty into the uncertainty related to the coefficient matrix A.

Specifically, one can put c into A with the addition of the constraint

$$z - c^T x \leq 0, \tag{7.10}$$

replacing the objective function (7.4) with

$$\max_{x,z} z \tag{7.11}$$

and adding z as another continuous variable.

One can put b into the A matrix by replacing (7.5) with

$$\sum_j a_{ij} x_j - b_i x_{n+1} \leq 0, \qquad \forall i, \tag{7.12}$$

and adding the constraint

$$x_{n+1} = 1. \tag{7.13}$$

The goal is to be robust against the worst-case scenario for a given level of conservatism. This level can be quantified using Γ_i, which represents the number of coefficients in constraint i that are allowed to vary in their uncertainty range. The other coefficients are assumed to have their nominal values, $\bar{a}_{ij}$. Let a_{ij} and δ_{ij} represent the specific elements of A and Δ, respectively. For the purposes of network optimization, we restrict Γ_i to integer values, but (Bertsimas and Sim 2004) details an extended robust program that allows for continuous choices of Γ_i. This robust problem is formally stated as

$$\min_I \left[\max_x \left(\sum_j c_j x_j \right) \right] \tag{7.14}$$

subject to

$$\sum_j a_{ij} x_j \leq b_i, \ \forall i, \tag{7.15}$$

$$a_{ij} \geq \bar{a}_{ij} - I_{ij}\delta_{ij}, \qquad \forall i, j, \tag{7.16}$$

$$a_{ij} \leq \bar{a}_{ij} + I_{ij}\delta_{ij}, \qquad \forall i, j, \tag{7.17}$$

$$\sum_j I_{ij} \leq \Gamma_i, \qquad \forall i, \tag{7.18}$$

$$\ell \leq x \leq u, \tag{7.19}$$

$$I_{ij} \in \{0, 1\}. \tag{7.20}$$

Here, Γ_i of the uncertain coefficients in constraint i are allowed to have a value other than $\bar{a}_{ij}$, and the choice of which specific coefficients have values other than their nominal value, $\bar{a}_{ij}$, is selected via the I_{ij} variables. The selection, according to (7.14), is those that decrease the objective function the most.

It would be excessively conservative to assume that all coefficients will deviate from their nominal values $\bar{a}_{ij}$. Instead the above formulation permits a more nuanced formulation reflecting one's level of conservatism for each constraint. The coefficients that are allowed to vary are indicated by the binary I_{ij} variables. Constraints (7.16) and (7.17) capture the coefficient variation chosen by the I_{ij} variables, and (7.18) places an upper bound on the number of variations in constraint i. By optimizing over the selection of I_{ij} variables in the objective function, the solution is feasible for the worst-case scenario of coefficient changes.

Following the approach of Bertsimas and Sim (2004), one solves (7.14)–(7.20) for the case of all $a_{ij} \geq 0$ by first replacing (7.15)–(7.17) with

$$\sum_j \bar{a}_{ij} x_j + B_i(x, \Gamma_i) \leq b_i, \qquad \forall i, \tag{7.21}$$

where

$$\begin{aligned} B_i(x, \Gamma_i) &= \max_{I_{i\cdot}} \left[\sum_j \delta_{ij} |x_j| I_{ij} \right] \text{ subject to } \sum_j I_{ij} \leq \Gamma_i \text{ and } 0 \leq I_{ij} \leq 1 \ \forall j, \\ &= \min_{p,z} \left[z_i \Gamma_i + \sum_j p_{ij} \right] \text{ subject to. } z_i + p_{ij} \geq \delta_{ij} |x_j| \ \forall \ j, \\ p_{ij} &\geq 0 \ \forall j, z_i \geq 0. \end{aligned} \tag{7.22}$$

The first optimization in (7.22) captures the intended impact of the I's. It will select a value for the I's to make the feasible region of (7.15) as small as possible as this will hinder the maximization with respect to x as much as possible. The second optimization is the dual of the first where z_i is the dual variable of the first constraint of the first optimization, and p_{ij} is the dual variable of the second. Since (7.21) is a "less than or equal to" constraint, the minimum in (7.22) will be achieved automatically in the quest to make the feasible region as large as possible. Since the solution of $\min y_j$ subject to $-y_j \leq x_j \leq -y_j$ is $y_j = |x_j|$, a similar argument can be applied to (7.25) and (7.26).

Combining these observations, solving (7.14)–(7.20) is equivalent to solving

$$\max_{x,y,z,p} \left[\sum_j c_j x_j \right] \tag{7.23}$$

subject to

$$\sum_j \bar{a}_{ij} x_j + z_i \Gamma_i + \sum_j p_{ij} \leq b_i, \qquad i = 1, ..., m, : \lambda \tag{7.24}$$

$$\delta_{ij} y_j - z_i - p_{ij} \leq 0, \qquad i = 1, ..., m; j = 1, ..., n, : \gamma \tag{7.25}$$

$$-y_j \leq x_j \leq y_j, \qquad j = 1, ..., n, : \phi \tag{7.26}$$

$$y, p, z \geq 0, \tag{7.27}$$

$$\ell \leq x \leq u. \tag{7.28}$$

The p, y, and z variables enforce the feasibility of the solution for the worst-case scenario of Γ_i. The dual variables λ, γ, and ϕ to the right of the linear program are displayed for ease of comparison with (7.38)–(7.43) below. One might note that if the number of coefficients that change is larger than Γ_i, (Bertsimas and Sim 2004) demonstrates that the robust solution will be feasible with a high probability.

7.2.2 Step 2: Constraining the Total Number of Coefficients not Equal to Their Nominal Value

Equations (7.23)–(7.28) present a complete solution for the robust optimization problem when there is a constraintwise bound on the number of coefficients that might change, Γ_i. For robust mission network analysis of interest in this book, a different but related robust optimization formulation is needed. Specifically, in this chapter we have a single parameter for the general level of conservatism, Γ_{tot}, in the system. The variable Γ_{tot} represents the integer number of uncertain coefficients in the linear program among *all* the constraints that are allowed to vary within their uncertainty range. This problem can be formulated with the following altered robust program:

$$\min_I \left[\max_x \left(\sum_j c_j x_j \right) \right] \tag{7.29}$$

subject to

$$\sum_j a_{ij} x_j \leq b_i, \qquad \forall i, \tag{7.30}$$

$$a_{ij} \geq \bar{a}_{ij} - I_{ij}\delta_{ij}, \qquad \forall i, j, \tag{7.31}$$

$$a_{ij} \leq \bar{a}_{ij} + I_{ij}\delta_{ij}, \qquad \forall i, j, \tag{7.32}$$

$$\sum_i \sum_j I_{ij} \leq \Gamma_{tot}, \tag{7.33}$$

$$\ell \leq x \leq u, \tag{7.34}$$

$$I_{ij} \in \{0, 1\}. \tag{7.35}$$

Note that the change from (7.18) to (7.33) allows the program to minimize over all Γ_i parameter variations subject to Γ_{tot}. The seemingly small change, from solving (7.14)–(7.20) to solving (7.29)–(7.35), uses Γ_{tot} to address the practical challenge of optimizing decisions given a desired level of hedging against uncertainty. However, this change requires an extension of Bertsimas and Sim (2004). As described in more detail below, this entails two steps:

- **Step 2a**: Find the dual of the robust primal (7.23)–(7.28).
- **Step 2b**: Change all the Γ_i to be variables rather than constants, and add the constraint $\sum_i \Gamma_i \leq \Gamma_{tot}$. The optimization yields the optimal values of Γ_i^*. While the optimal values, Γ_i^*, might initially appear to be the end result, they are intermediate results, which are required to determine the optimal decisions, x, in (7.14)–(7.20).

Step 2a: The key step is to find the dual of (7.23)–(7.28) after replacing (7.26) with

$$-x_j - y_j \leq 0, \qquad j = 1, ..., n : \phi_1, \tag{7.36}$$

and

$$x_j - y_j \leq 0, \qquad j = 1, ..., n : \phi_2. \tag{7.37}$$

The dual is (7.38)–(7.43) using the variables: $\lambda \in \mathcal{R}^m$, $\gamma \in \mathcal{R}^{m \times n}$, and $\phi \in \mathcal{R}^{2 \times n}$. For convenience of comparison, the dual variables are listed next to the corresponding constraints in the robust primal, and the primal variables are listed next to the corresponding constraints in the robust dual.

$$\min_{\Gamma}\left[\sum_{i=1}^{m} b_i\lambda_i\right] \tag{7.38}$$

subject to

$$\sum_{i=1}^{m} \bar{a}_{ij}\lambda_i - \phi_{1j} + \phi_{2j} \geq c_j, \qquad j = 1, ..., n, : x, \tag{7.39}$$

$$\sum_{i=1}^{m} \delta_{ij}\gamma_{ij} - \phi_{1j} - \phi_{2j} \geq 0, \qquad j = 1, ..., n, : y, \tag{7.40}$$

$$\Gamma_i\lambda_i - \sum_{j=1}^{n} \gamma_{ij} \geq 0, \qquad i = 1, ..., m, : z, \tag{7.41}$$

$$\lambda_i - \gamma_{ij} \geq 0, \qquad \forall i = 1, ..., m; j = 1, ..., n, : p, \tag{7.42}$$

$$\lambda, \phi, \gamma \geq 0. \tag{7.43}$$

Step 2b: Having found the dual, one should now view the Γ_i's as variables and add the constraint

$$\sum_i \Gamma_i \leq \Gamma_{tot}. \tag{7.44}$$

Allowing Γ_i to vary while computing the optimization in (7.38)–(7.44) allows the program to evaluate the most diminishing combination of changed coefficients against which we must be robust. Specifically, the optimal variables, Γ_i^*, found are put into (7.14)–(7.20) as constants, and this primal optimization is solved. This solution will be robust against the worst-case scenario within the uncertainty for a given value of Γ_{tot}.

7.2.3 Step 3: Linearizing the New Formulation

Unfortunately, (7.41) contains the nonlinear term $\Gamma_i\lambda_i$. This is undesirable as the computation time of solving a nonlinear optimization is considerably longer than solving a linear optimization. Fortunately, it is possible to linearize this quadratic $\Gamma_i\lambda_i$ term with the addition of the binary variable ξ_{ij} and the variable Ξ_{ij} using a simplification of the McCormick Envelopes described in McCormick (1976).

As explained below, it is shown that the formulation (7.46)–(7.57) is both linear and equivalent to (7.38)–(7.43). The reasoning entails two steps which will be explained below:

- **Step 3a**: Eqs. (7.53)–(7.57) imply

$$\Xi_{ij} = \begin{cases} 0, & \text{if } \xi_{ij} = 0, \\ \lambda_i, & \text{if } \xi_{ij} = 1. \end{cases} \tag{7.45}$$

- **Step 3b**: Eqs. (7.52) and (7.45) imply that (7.49) is equivalent to (7.41).

$$\min_{\Gamma} \left[\sum_{i=1}^{m} b_i \lambda_i \right] \tag{7.46}$$

subject to

$$\sum_{i=1}^{m} \bar{a}_{ij} \lambda_i - \phi_{1j} + \phi_{2j} \ge c_j, \qquad j = 1, ..., n, \tag{7.47}$$

$$\sum_{i=1}^{m} \delta_{ij} \gamma_{ij} - \phi_{1j} - \phi_{2j} \ge 0, \qquad j = 1, ..., n, \tag{7.48}$$

$$\sum_{j=1}^{n} \left(\Xi_{ij} - \gamma_{ij} \right) \ge 0, \qquad i = 1, ..., m, \tag{7.49}$$

$$\lambda_i - \gamma_{ij} \ge 0, \qquad i = 1, ..., m;\ j = 1, ..., n, \tag{7.50}$$

$$\sum_{i=1}^{m} \Gamma_i \le \Gamma_{tot}, \tag{7.51}$$

$$\sum_{j=1}^{n} \xi_{ij} = \Gamma_i, \qquad i = 1, ..., m, \tag{7.52}$$

$$\Xi_{ij} \le \xi_{ij} M, \qquad i = 1, ..., m;\ j = 1, ..., n, \tag{7.53}$$

$$\Xi_{ij} \le \lambda_i, \qquad i = 1, ..., m;\ j = 1, ..., n, \tag{7.54}$$

$$\Xi_{ij} \ge \lambda_i - (1 - \xi_{ij}) M, \qquad i = 1, ..., m;\ j = 1, ..., n, \tag{7.55}$$

$$\lambda, \phi, \gamma, \Gamma, \Xi \ge 0, \tag{7.56}$$

$$\xi \in \{0, 1\}. \tag{7.57}$$

Step 3a: To see that Eqs. (7.53)–(7.57) imply (7.45), one must verify it for the following two cases:

Case 1, $\xi_{ij} = 0$: In this case, (7.55) has no impact on Ξ_{ij} as M is very large. However, (7.53), (7.56) (and (7.54)) imply $\Xi_{ij} = 0$ as stated in (7.45).

Case 2, $\xi_{ij} = 1$: Here, (7.53) has no impact on Ξ_{ij} as M is very large. However, (7.54) and (7.55) imply $\Xi_{ij} = \lambda_i$ as stated in (7.45).

Step 3b: To see that Eqs. (7.52) and (7.45) imply that (7.49) is equivalent to (7.41), one first notes that from (7.45) and (7.52), the right-hand side of (7.49), $\sum_{j=1}^{n} \left(\Xi_{ij} - \gamma_{ij}\right) = \lambda_i \Gamma_i - \sum_{j=1}^{n} \gamma_{ij}$, which is indeed the right-hand side of (7.41).

7.3 Illustrative Example

7.3.1 Procedure and Comparison

To illustrate the analysis of the previous section, consider the optimization (7.14)–(7.20) along with the constraint $\Gamma_1 + \Gamma_2 \le \Gamma_{tot}$, where

$$\bar{A} = \begin{bmatrix} 1 & 2 & 3 \\ 5 & 4 & 3 \end{bmatrix}, \ \Delta = \begin{bmatrix} 1 & 0.4 & 0.6 \\ 0.2 & 3 & 1.1 \end{bmatrix}, \ b = \begin{bmatrix} 5 \\ 12 \end{bmatrix}, \ c = \begin{bmatrix} 5 \\ 5 \\ 5 \end{bmatrix}, \ \text{and } x = \begin{bmatrix} x_1 \\ x_2 \\ x_3 \end{bmatrix},$$

and where $\bar{a}_{ij}$ and δ_{ij} are the (i, j)th component of $\bar{A}$ and Δ, respectively, and b_i, c_i, and x_i are the ith components of b, c, and x, respectively. Table 7.1 displays the optimal solution for the six alternative values of Γ_{tot} and identifies the optimal values of Γ_1, Γ_2, x, and the objective function (7.14). This is done in two steps: First, (7.46)–(7.57) are used to identify the optimal values of Γ_1, and Γ_2. Next, these values are used to find the optimal values of x and the objective function by solving (7.23)–(7.28) and using Γ_1 and Γ_2 as known constants.

In the absence of the approach of this chapter, one would be forced to use a computationally inefficient alternative to produce Table 7.1 by examining every combination of $[\Gamma_1, \Gamma_2]$ using (7.14)–(7.20) and treating Γ_1 and Γ_2 as given constants. The output for every combination is reported in Table 7.2. The diagonals (bottom left to top right) correspond to constant Γ_{tot} values with the smallest objective along those diagonals bolded. The combinations associated with those bold entries are the optimal robust combinations at that level of Γ_{tot}.

An examination of Tables 7.1 and 7.2 shows consistent results, although Table 7.1 was more efficient to generate. For example, suppose $\Gamma_{tot} = 2$. The three options for $[\Gamma_1, \Gamma_2]$ are [2,0], [1,1], and [0,2], which correspond to objective values 12.25, 12.067, and 12.876, respectively. The minimum of those objective values is 12.067, which is consistent with Table 7.1. The bolded value corresponds to $[\Gamma_1 = 1, \Gamma_2 = 1]$ and is the scenario against which we must be robust.

For this small example, the computational efficiency savings are of minimal consequence. However, for larger examples, the impact is substantial as can be seen as follows: Suppose $A \in \mathcal{R}^{m \times n}$. There are m constraints in A, so there are m different constraintwise variables, Γ_i. Each such variable can have $n + 1$ different values $(0, 1, \ldots, n)$, since each constraint has n coefficients. Hence the full enumeration approach of Table 7.2 requires $(n + 1)^m$ optimizations to solve for all values of Γ_{tot}. In contrast, since A has mn coefficients, there are $mn + 1$ different

Table 7.1 Full output from the new approach

Γ_{tot}	$[\Gamma_1, \Gamma_2]$	Objective	$[x_1, x_2, x_3]$
0	[0,0]	14.167	[0.67, 2.17, 0.00]
1	[1,0]	12.500	[0.71, 1.78, 0.00]
2	[1,1]	12.067	[2.10, 0.14, 0.17]
3	[1,2]	11.896	[2.05, 0.09, 0.24]
4	[2,2]	11.798	[2.12, 0.06, 0.18]
5	[2,3]	11.740	[2.16, 0.00, 0.19]
6	[3,3]	11.740	[2.16, 0.00, 0.19]

Table 7.2 Output from an exhaustive search using (7.14)–(7.20)

		Γ_2			
		0	1	2	3
Γ_1	0	**14.167**	13.257	12.876	12.826
	1	**12.5**	**12.067**	**11.896**	11.824
	2	12.25	11.978	**11.798**	**11.740**
	3	12.25	11.929	11.782	**11.740**

values of Γ_{tot}. To construct Table 7.1, for each value of Γ_{tot} only two optimizations are needed: one to find the optimal values of $\Gamma_1, \Gamma_2, \ldots, \Gamma_m$ using (7.46)–(7.57) and one using these values to find the optimal decisions using (7.23)–(7.28). Hence, to solve for all values of Γ_{tot}, the new approach only requires solving $2(mn + 1)$ optimizations, whereas the full enumeration approach requires solving $(n + 1)^m$ optimization.

As a side note, one might observe that the minimax formulation in (7.29) and Table 7.2 can be thought of as a two-player, two-stage game, sometimes referred to as a Stackelberg Game, where nature gets to "choose" the $[\Gamma_1, \Gamma_2]$ for a given level of Γ_{tot}. Then the user maximizes $\sum_j c_j x_j$ by choosing a feasible x given the worst-case scenario allowed by the chosen $[\Gamma_1, \Gamma_2]$ combination.

7.3.2 *Choosing the Best Level of Conservatism,* Γ_{tot}

With a sufficiently large linear program, it is easier to visualize the effects of increased levels of conservatism with a plot such as Fig. 7.1 to, in the words of Bertsimas and Sim (2004), quantify the price of robustness. Figure 7.1 shows the results of a randomly generated linear programs with 50 variables and 50 constraints. The plot is limited to $\Gamma_{tot} \leq 50$ because, beyond that, the change in objective is virtually 0. The input needed to create this plot with the new robust approach required solving 2,501 linear programs and 2,499 mixed integer programs. To produce this input using the (Bertsimas and Sim 2004) approach for every possible combination of Γ_i, the user would have to solve $51^{50} \approx 10^{84}$ different

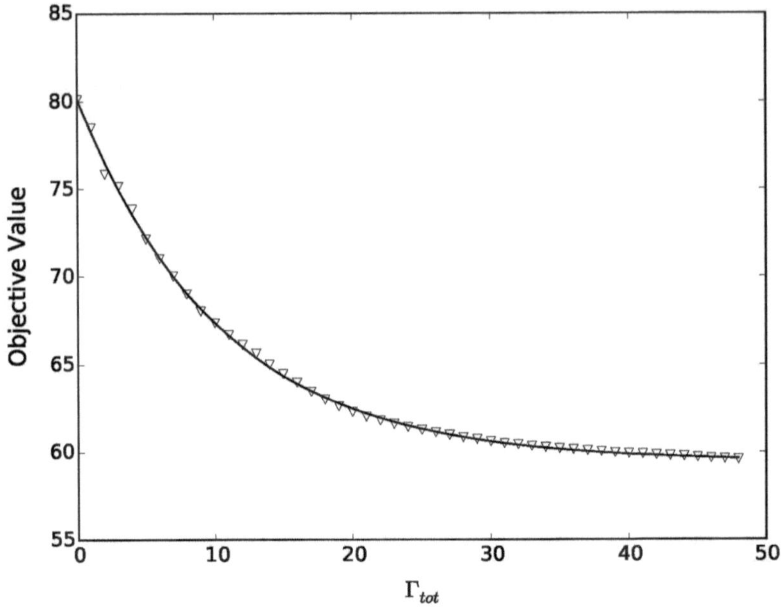

Fig. 7.1 The exponential decay of the objective function of a randomly generated linear program as the conservatism parameter Γ_{tot} is increased. This linear program has 50 variables and 50 constraints. We fit an exponential curve of the form: $\hat{y} = ae^{bx} + c$

linear programs. Even if the user knew a priori that Γ_{tot} could be limited to 50, the number of combinations would be $C(99, 49) = 99!/(49!50!) \approx 10^{28}$ linear programs (the number of combinations is found by recognizing that this is a classical Star and Bar Problem with 99 stars and 49 bins) (Feller 1968).

Figure 7.1 allows the user to fully evaluate the consequences of each level of Γ_{tot} quickly and requires no fine-tuned model decisions. An exponential curve is fit to the data. For the linear programs examined, it was frequently the case that a near-exponential curve, except in the tail, emerged. Figure 7.2 visualizes the same data as Fig. 7.1 on a log scale with the fitted asymptote removed. The r-squared value for this linear fit is 0.997, but it can be seen that the tail diverges from the fit and accelerates downward. This sub-exponential behavior was present in all of the linear programs tested.

Due to the ubiquity of this simple form, one might posit that the resiliency of a linear program, or a mission network in particular, against uncertainty can be characterized by the slope of the fitted line. The slope captures the sensitivity of the mission objective to the number of coefficients, which deviate from their assumed nominal values. Hence, one might intuit that a steeper slope is associated with the less resilient system and there the slope might be a valuable proxy for resiliency.

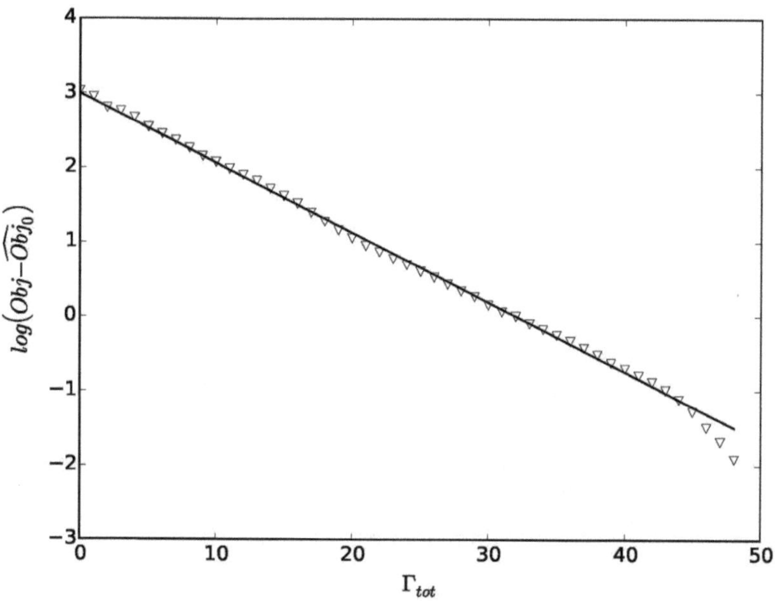

Fig. 7.2 The log of the objective values less the fitted asymptote from the exponential curve in Fig. 7.1. The objective values vary exponentially with the conservatism parameter Γ_{tot} except for the tail. Divergence in the tail is typical in the linear programs we have studied

7.4 Robust Mission Defense

7.4.1 The Context

FDNA, first introduced in the literature with Garvey and Pinto (2009) and introduced in Chap. 1 of this book, is a dependency representation that quantifies the propagation of the operability of component nodes within a functional mission network. The dependency relationships of each link are characterized by two parameters, α and β, which are implicitly defined by (7.58). The operability value of node i, P_i, is its MOE. The mission operability for the network is equal to the MOE of the top node in the network, i.e., the only node that depends on all other nodes directly and indirectly. The robust defensive mission network analysis below assumes, for an illustrative purpose, a mission network description using FDNA and implicitly points the direction for the required analysis when using non-FDNA frameworks.

In the context of mission defense, there are the player, BLUE, who defends the network, and an adversary, RED, who attacks the network. The MOE of each leaf node is determined based on the level of defense by BLUE of the node, under the assumption that a low level of defense creates an exploitable opportunity for RED, and a high level of defense reduces the exploitable opportunity for RED, though not

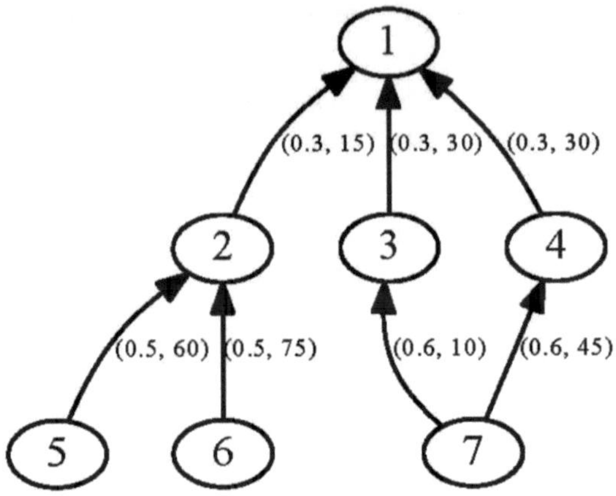

Fig. 7.3 The small network used to demonstrate Mission Defense Analysis using FDNA. The edge labels are the corresponding (α, β) parameters. Note that the arrows flow from child node to parent node corresponding to the direction of value propagation through the network. The nominal operating levels for leaf nodes 5, 6, and 7 are 45, 25, and 80, respectively

necessarily completely. For this robust mission defense example, the values of α and β are only known imperfectly. We assume that their true values α and β are in a fixed range denoted by $[\bar{\alpha} - \delta^{\alpha}, \bar{\alpha} + \delta^{\alpha}]$ and $\left[\bar{\beta} - \delta^{\beta}, \bar{\beta} + \delta^{\beta}\right]$, respectively. The analysis in the preceding sections will be applied to this scenario to find the robust optimal defense strategy, given that only Γ_{tot} of the dependency parameters deviate from their nominal values, $\bar{\alpha}$ and $\bar{\beta}$. The variable Γ_{tot} might be interpreted as the risk adversity of BLUE.

The robust mission defense analysis approach below consists of first introducing notation in Sect. 7.4.2 and framing its non-uncertainty analysis variation in Sect. 7.4.3. Next, the general approach of Step 1, Step 2, and Step 3 of the analysis presented in Sects. 7.2.1, 7.2.2, and 7.2.3, respectively, is specialized to FDNA below in Sects. 7.4.4, 7.4.5, and 7.4.6, respectively. Finally, in Sect. 7.4.7, for illustrative purposes, the results are further specialized to the example displayed in Fig. 7.3.

7.4.2 Notation

The formal formulation of the robust mission defense example uses the following notation:

$w_i \in [0, 1]$ is BLUE's defense investments in node i where $w_i = 1$ corresponds to BLUE fully defending node i.

$\gamma_i \in [0, 100]$ is the nominal operability of node i in the absence of an attack.

c_i is the direct cost of BLUE's defense investment in node i.

$P_i \in [0, 100]$ is the MOE or the operability value of node i. The MOE would typically be a concave function of the Measure of Performance (MOP) (e.g., widget/sec) of the node.

$P_1 \in [0, 100]$ is the MOE or operability value of the top node, node 1, which is also the node corresponding to the MOE of the mission.

$\mathcal{A}$ is the set of all nodes.

$\mathcal{L}$ is the set of leaf nodes. These are nodes whose MOE is not dependent on other nodes.

$\mathcal{N}$ is the set of non-leaf nodes.

n is the number of non-leaf nodes: $|\mathcal{N}|$.

$\mathcal{J}(m)$ is the set of nodes that directly impact the MOE of node m. This will be referred to as the *children* of node m. For this formulation, it is assumed that the actions of RED will impact the *leaf nodes* of the network, i.e., nodes with the property that $\mathcal{J}(m) = \{\}$.

$\mathcal{K}(m)$ is the set of the *parents* of node m, i.e., if $i \in \mathcal{K}(m)$, then $m \in \mathcal{J}(i)$.

α and β are the edge parameters which are unknown.

$\bar{\alpha}, \bar{\beta}$ are the nominal values of the edge parameters.

δ^α, δ^β are the widths of the uncertainty of the edge parameters, i.e., $\alpha \in [\bar{\alpha} - \delta^\alpha, \bar{\alpha} + \delta^\alpha]$ and $\beta \in \left[\bar{\beta} - \delta^\beta, \bar{\beta} + \delta^\beta\right]$.

7.4.3 *Step 0: The Defense Mission Analysis Primal Formulation in Absence of Uncertainty*

Equation (7.58) characterizes a parent node's operability in terms of the operabilities of its dependent child nodes assuming FDNA dependencies used throughout this book (Garvey and Pinto 2009; Servi and Garvey 2017). It is a recursive formula which ultimately relates the operability of each node to the operability of the leaf nodes. The leaf nodes' operabilities are assumed to be based on their nominal operability values and the defense investments of BLUE.

$$P_i = \min\left[100 - \frac{1}{|\mathcal{J}(i)|}\sum_{j\in\mathcal{J}(i)}[\alpha_{ij}\left(100 - P_j\right)],\ \min_{j\in\mathcal{J}(i)}\left[P_j + \beta_{ij}\right]\right]. \tag{7.58}$$

The defense mission analysis primal formulation in the absence of uncertainty is as follows:

$$\max_{w}\ P_1 \tag{7.59}$$

subject to

$$P_i + \frac{1}{|\mathcal{J}(i)|}\sum_{j\in\mathcal{J}(i)}\left[\alpha_{ij}\left(100 - P_j\right)\right] \le 100, \qquad \forall i \in \mathcal{N}, \tag{7.60}$$

$$P_i - P_j \le \beta_{ij}, \qquad \forall i \in \mathcal{N}, j \in \mathcal{J}(i), \tag{7.61}$$

$$P_i - 0.6 w_i \gamma_i \le 0.3\gamma_i, \qquad \forall i \in \mathcal{L}, \tag{7.62}$$

$$\sum_i w_i c_i \le B, \tag{7.63}$$

$$P_i \le 100, \qquad \forall i \in \mathcal{A}, \tag{7.64}$$

$$w_i \le 1, \qquad \forall i \in \mathcal{L}, \tag{7.65}$$

$$w_i \ge 0, \qquad \forall i \in \mathcal{L}, \tag{7.66}$$

$$P_i \ge 0, \qquad \forall i \in \mathcal{A}. \tag{7.67}$$

In the absence of uncertainty in the α and β parameters, the optimal BLUE defense investment decision is the solution to (7.59)–(7.67). BLUE seeks to maximize MOE of the mission, P_1, (7.59). The FDNA equation (7.58) is captured in (7.60) and (7.61). The budget of BLUE's defense investments is enforced by (7.63). Leaf nodes do not have children, so their values are calculated differently according to the level of BLUE's defense investment. It will be assumed that BLUE has the ability to minimize the degradation of a leaf node's operability from RED to only 90% of its nominal value by maximally investing in its defense, i.e., choosing $w_i = 1$, but leaving that leaf node's operability vulnerable to being degraded by RED to 30% of its nominal value by RED if not defended, i.e., choosing $w_i = 0$. This is captured in (7.62).

7.4.4 Step 1: Constraining the Number of Coefficients not Equal to Their Nominal Value for Each Constraint

To add robustness to this optimization with respect to the α and β parameters, start by modifying (7.61) in the manner described in Sect. 7.2.1, where (7.5) was replaced by (7.12) and (7.13). Specifically, because β_{ij} is on the right-hand side of (7.61), first replace (7.61) with

$$P_i - P_j - \beta_{ij} b_0 \le 0, \qquad \forall i \in \mathcal{N}, j \in \mathcal{J}(i), \tag{7.68}$$

and

$$b_0 = 1. \tag{7.69}$$

After this replacement, one can proceed with finding the robust optimal mission defense.

We transform the nominal formulation (7.59)–(7.60) and (7.62)–(7.69) to the robust formulation (7.70)–(7.86) the same way that we transformed (7.4)–(7.6) into (7.23)–(7.28). This is summarized in Table 7.3. Specifically, (7.71) and (7.72) correspond to (7.24) when applied to (7.60) and (7.68); Eqs. (7.76) and (7.77)

Table 7.3 Summary of equations used to transform formulation to the defense mission analysis optimization

Equation	Previous equation	Guiding equation
(7.71)	(7.60)	(7.24)
(7.72)	(7.68)	(7.24)
(7.73)	(7.62)	–
(7.74)	(7.63)	–
(7.75)	(7.69)	–
(7.76)	(7.60)	(7.25)
(7.77)	(7.68)	(7.25)
(7.78), (7.79)	(7.26)	–
(7.80), (7.81)	(7.26)	–
(7.82)	(7.64)	–
(7.83)	(7.65)	–

correspond to (7.25) when applied to (7.60) and (7.68); Eqs. (7.78) and (7.79) correspond to (7.26), where $100 - P_j$ is used rather than P_j as that is the coefficient of α_i in (7.60); Eqs. (7.80) and (7.81) correspond to (7.26); and finally, the equations not containing α's or β's, (7.73), (7.74), (7.75), (7.82), and (7.83), are simply (7.62), (7.63), (7.69), (7.64), and (7.65), respectively.

$$\max_{w} \ P_1 \tag{7.70}$$

subject to

$$P_i + \frac{1}{|\mathcal{J}(i)|} \sum_{j \in \mathcal{J}(i)} \left[\alpha_{ij}\left(100 - P_j\right)\right] + z_i^{\alpha}\Gamma_i^{\alpha} + \sum_{j \in \mathcal{J}(i)} p_{ij}^{\alpha} \leq 100, \quad \forall i \in \mathcal{N}, \quad : \lambda^{\alpha} \tag{7.71}$$

$$P_i - P_j - \beta_{ij} b_0 + z_{ij}^{\beta}\Gamma_{ij}^{\beta} + p_{ij}^{\beta} \leq 0, \qquad \forall i \in \mathcal{N}, j \in \mathcal{J}(i), \qquad : \lambda^{\beta} \tag{7.72}$$

$$P_i - 0.6 w_i \gamma_i \leq 0.3\gamma_i, \qquad \forall i \in \mathcal{L}, \qquad : \lambda^{\gamma} \tag{7.73}$$

$$\sum_i w_i c_i \leq B, \qquad : \mu \tag{7.74}$$

$$b_0 = 1, \qquad : \varepsilon \tag{7.75}$$

$$\frac{\delta_{ij}^{\alpha} y_j}{|\mathcal{J}(i)|} - z_i^{\alpha} - p_{ij}^{\alpha} \leq 0, \qquad \forall i \in \mathcal{N}, j \in \mathcal{J}(i), \qquad : \phi^{\alpha} \tag{7.76}$$

$$\delta_{ij}^{\beta} y_0 - z_{ij}^{\beta} - p_{ij}^{\beta} \leq 0, \qquad \forall i \in \mathcal{N}, j \in \mathcal{J}(i), \qquad : \phi^{\beta} \tag{7.77}$$

$$(100 - P_j) - y_j \leq 0, \qquad \forall j \in \mathcal{A}, j \neq 1, \qquad : \eta_{j1} \tag{7.78}$$

$$-y_j - (100 - P_j) \leq 0, \qquad \forall j \in \mathcal{A}, i \neq 1, \qquad : \eta_{j2} \tag{7.79}$$

$$-y_0 - b_0 \leq 0, \qquad : \eta_{01} \tag{7.80}$$

$$b_0 - y_0 \leq 0, \qquad : \eta_{02} \tag{7.81}$$

$$P_i \leq 100, \qquad \forall i \in \mathcal{A}, \qquad : \delta^P \tag{7.82}$$

$$w_i \leq 1, \qquad \forall i \in \mathcal{L}, \qquad : \delta^w \tag{7.83}$$

$$P_i, w_i, y_i > 0, \qquad \forall i, \tag{7.84}$$

$$z_i^\alpha, z_{ij}^\beta \geq 0, \qquad \forall i \in \mathcal{N}, j \in \mathcal{J}(i), \tag{7.85}$$

$$p_{ij}^\alpha, p_{ij}^\beta \geq 0, \qquad \forall i \in \mathcal{N}, j \in \mathcal{J}(i). \tag{7.86}$$

7.4.5 Step 2: Constraining the Total Number of Coefficients not Equal to Their Nominal Value

The robust formulation (7.70)–(7.86) solves the problem of determining the best level of defensive node investments subject to a budget constraint, while also hedging against some parameters deviating from their assumed nominal values. The level of hedging is set using the variables: Γ_i^α and Γ_i^β. Γ_i^α corresponds to the number of α parameters feeding into parent node i that are not equal to their nominal values but instead may have any value in $[\alpha_{ij} - \delta_{ij}^\alpha, \alpha_{ij} + \delta_{ij}^\alpha]$. Γ_{ij}^β corresponds to the number of the β_{ij} parameters related to the (i, j) link that are not equal to their nominal values but instead could have any value in $[\beta_{ij} - \delta_{ij}^\beta, \beta_{ij} + \delta_{ij}^\beta]$.

However, as indicated in (7.60) and (7.61), there is only one β value associated with the link (i, j), and a defender is not typically concerned about the uncertainty of a value of α for a specific parent node but all of the parent nodes. This accentuates the need to extend the work of Bertsimas and Sim (2004). In this section, following the work in Sect. 7.2.2, the analysis will be extended to the assumption that up to Γ_{tot}^α values of α parameter of *any* parent node and up to Γ_{tot}^β values of β parameter of *any* link might not be equal to their nominal values. Once this solution is found, it is a simple generalization to solve the optimization for the case of up to Γ_{tot} values of any α or β parameter that might not be equal to their nominal values.

Following the approach of Sect. 7.2.2, the first step of the analysis is to identify the dual of Eqs. (7.70)–(7.86). It is given by

$$\min_{\Gamma} \Bigg[\sum_{i \in \mathcal{N}} \Bigg[100\lambda_i^\alpha \Bigg(1 - \frac{1}{|\mathcal{J}(i)|} \Bigg(\sum_{j \in \mathcal{J}(i)} \alpha_{ij} \Bigg) \Bigg) \Bigg] + \sum_{i \in \mathcal{L}} \Big[0.3\lambda_i^\gamma \gamma_i + \delta_i^w \Big] \tag{7.87}$$

$$+ 100 \sum_{i \in \mathcal{A}} \delta_i^P + B\mu + \varepsilon - 100 \sum_{i \in \mathcal{A}, i \neq 1} (\eta_{i1} - \eta_{i2}) \Bigg],$$

subject to

$$\lambda_1^\alpha + \sum_{j\in\mathcal{J}} \lambda_{1j}^\beta + \delta_1^P \geq 1, \qquad : P_1, \tag{7.88}$$

$$\lambda_i^\alpha + \sum_{j\in\mathcal{J}(i)} \lambda_{ij}^\beta - \sum_{k\in\mathcal{K}(i)} \left[\frac{\alpha_{ki}}{|\mathcal{J}(k)|}\lambda_k^\alpha + \lambda_{ki}^\beta\right] + \delta_i^P - \eta_{i1} + \eta_{i2} \geq 0,$$

$$\forall i \in \mathcal{N}, i \neq 1 : P_i, \tag{7.89}$$

$$\lambda_i^\gamma - \sum_{k\in\mathcal{K}(i)} \left[\frac{\alpha_{ki}}{|\mathcal{J}(k)|}\lambda_k^\alpha + \lambda_{ki}^\beta\right] + \delta_i^P - \eta_{i1} + \eta_{i2} \geq 0, \qquad \forall i \in \mathcal{L} : P_i, \tag{7.90}$$

$$\varepsilon - \sum_{i\in\mathcal{N}, j\in\mathcal{J}(i)} \left[\lambda_{ij}^\beta \beta_{ij}\right] - \eta_{01} + \eta_{02} \geq 0, \qquad : b_0, \tag{7.91}$$

$$-0.6\lambda_i^\gamma \gamma_i + \delta_i^w + c_i\mu \geq 0, \qquad \forall i \in \mathcal{L} : w, \tag{7.92}$$

$$\sum_{k\in\mathcal{K}(i)} \left[\frac{\phi_{ki}^\alpha \delta_{ki}^\alpha}{|\mathcal{J}(k)|}\right] - \eta_{i1} - \eta_{i2} \geq 0, \qquad \forall i \in \mathcal{A}, i \neq 1 : y, \tag{7.93}$$

$$\sum_{i\in\mathcal{N}, j\in\mathcal{J}(i)} \left[\phi_{ij}^\beta \delta_{ij}^\beta\right] - \eta_{01} - \eta_{02} \geq 0, \qquad : y_0, \tag{7.94}$$

$$\Gamma_i^\alpha \lambda_i^\alpha - \sum_{j\in\mathcal{J}(i)} \phi_{ij}^\alpha \geq 0, \qquad \forall i \in \mathcal{N} : z^\alpha, \tag{7.95}$$

$$\Gamma_{ij}^\beta \lambda_{ij}^\beta - \phi_{ij}^\beta \geq 0, \quad \forall i \in \mathcal{N}, j \in \mathcal{J}(i) : z^\beta, \tag{7.96}$$

$$\lambda_i^\alpha - \phi_{ij}^\alpha \geq 0, \quad \forall i \in \mathcal{N}, j \in \mathcal{J}(i) : p^\alpha, \tag{7.97}$$

$$\lambda_{ij}^\beta - \phi_{ij}^\beta \geq 0, \quad \forall i \in \mathcal{N}, j \in \mathcal{J}(i) : p^\beta, \tag{7.98}$$

$$\sum_{i\in\mathcal{N}} \Gamma_i^\alpha \leq \Gamma_{tot}^\alpha, \tag{7.99}$$

$$\sum_{i\in\mathcal{N}, j\in\mathcal{J}} \Gamma_{ij}^\beta \leq \Gamma_{tot}^\beta, \tag{7.100}$$

$$\Gamma^{\alpha}_{tot} + \Gamma^{\beta}_{tot} \leq \Gamma_{tot}, \tag{7.101}$$

$$\lambda, \mu, \phi, \eta, \delta, \Gamma \geq 0, \tag{7.102}$$

$$\psi, \varepsilon \qquad \text{unrestricted.} \tag{7.103}$$

The second step of the analysis is to view Γ^{α}_{i} and Γ^{β}_{ij} as variables. However, as found in Sect. 7.2.2, this creates nonlinearity in the formulation that must be addressed.

7.4.6 Step 3: Linearizing the New Formulation

As indicated at the end of Sect. 7.4.5, the final step is to remove the nonlinearity in the optimization formulation in order to improve computation time. This is done by applying the approach of Sect. 7.2.3 to Eqs. (7.87)–(7.103). In linearizing this particular example, note that we add ξ^{α}, but not ξ^{β}. The reason for this is that Γ^{β}_{ij} is already binary by the nature of the problem; as such we do not need to add another binary variable for each edge for the β parameter. For the sake of maintaining intuition in the formulation, we linearize Γ^{α}_{tot} and Γ^{β}_{tot} separately. The formulation for the linearization is presented as follows:

$$\min_{\Gamma} \left[\sum_{i \in \mathcal{N}} \left[100\lambda^{\alpha}_{i} \left(1 - \frac{1}{|\mathcal{J}(i)|} \left(\sum_{j \in \mathcal{J}(i)} \alpha_{ij} \right) \right) \right] + \sum_{i \in \mathcal{L}} \left[0.3\lambda^{\gamma}_{i} \gamma_i + \delta^{w}_{i} \right] \right.$$
$$\left. +100 \sum_{i \in \mathcal{A}} \delta^{P}_{i} + B\mu + \varepsilon - 100 \sum_{i \in \mathcal{A}, i \neq 1} (\eta_{i1} - \eta_{i2}) \right], \tag{7.104}$$

subject to

$$\lambda^{\alpha}_{1} + \sum_{j \in \mathcal{J}} \lambda^{\beta}_{1j} + \delta^{P}_{1} \geq 1, \qquad : P_1, \tag{7.105}$$

$$\lambda^{\alpha}_{i} + \sum_{j \in \mathcal{J}(i)} \lambda^{\beta}_{ij} - \sum_{k \in \mathcal{K}(i)} \left[\frac{\alpha_{ki}}{|\mathcal{J}(k)|} \lambda^{\alpha}_{k} + \lambda^{\beta}_{ki} \right] + \delta^{P}_{i} - \eta_{i1} + \eta_{i2} \geq 0,$$
$$\forall i \in \mathcal{N}, i \neq 1 : P_i, \tag{7.106}$$

$$\lambda^{\gamma}_{i} - \sum_{k \in \mathcal{K}(i)} \left[\frac{\alpha_{ki}}{|\mathcal{J}(k)|} \lambda^{\alpha}_{k} + \lambda^{\beta}_{ki} \right] + \delta^{P}_{i} - \eta_{i1} + \eta_{i2} \geq 0, \qquad \forall i \in \mathcal{L}, : P_i, \tag{7.107}$$

$$\varepsilon - \sum_{i\in\mathcal{N}, j\in\mathcal{J}(i)} [\lambda_{ij}^{\beta}\beta_{ij}] - \eta_{01} + \eta_{02} \geq 0, \qquad : b_0, \tag{7.108}$$

$$-0.6\lambda_i^{\gamma}\gamma_i + \delta_i^{w} + c_i\mu \geq 0, \qquad \forall i \in \mathcal{L}, : w, \tag{7.109}$$

$$\sum_{k\in\mathcal{K}(i)} \left[\frac{\phi_{ki}^{\alpha}\delta_{ki}^{\alpha}}{|\mathcal{J}(k)|}\right] - \eta_{i1} - \eta_{i2} \geq 0, \qquad \forall i \in \mathcal{A}, i \neq 1 : y, \tag{7.110}$$

$$\sum_{i\in\mathcal{N}, j\in\mathcal{J}(i)} [\phi_{ij}^{\beta}\delta_{ij}^{\beta}] - \eta_{01} - \eta_{02} \geq 0, \qquad : y_0, \tag{7.111}$$

$$\sum_{j\in\mathcal{J}(i)} [\Xi_{ij}^{\alpha} - \phi_{ij}^{\alpha}] \geq 0, \qquad \forall i \in \mathcal{N}, \tag{7.112}$$

$$\Xi_{ij}^{\beta} - \phi_{ij}^{\beta} \geq 0, \qquad \forall i \in \mathcal{N}, j \in \mathcal{J}(i), \tag{7.113}$$

$$\lambda_i^{\alpha} - \phi_{ij}^{\alpha} \geq 0, \qquad \forall i \in \mathcal{N}, j \in \mathcal{J}(i), \tag{7.114}$$

$$\lambda_{ij}^{\beta} - \phi_{ij}^{\beta} \geq 0, \qquad \forall i \in \mathcal{N}, j \in \mathcal{J}(i), \tag{7.115}$$

$$\sum_{i\in\mathcal{N}, j\in\mathcal{J}} \xi_{ij}^{\alpha} \leq \Gamma_{tot}^{\alpha}, \tag{7.116}$$

$$\sum_{i\in\mathcal{N}, j\in\mathcal{J}} \Gamma_{ij}^{\beta} \leq \Gamma_{tot}^{\beta}, \tag{7.117}$$

$$\Gamma_{tot}^{\alpha} + \Gamma_{tot}^{\beta} \leq \Gamma_{tot}, \tag{7.118}$$

$$\Xi_{ij}^{\alpha} \leq \xi_{ij}^{\alpha} M, \qquad \forall i \in \mathcal{N}, j \in \mathcal{J}(i), \tag{7.119}$$

$$\Xi_{ij}^{\alpha} \leq \lambda_i^{\alpha}, \qquad \forall i \in \mathcal{N}, j \in \mathcal{J}(i), \tag{7.120}$$

$$\Xi_{ij}^{\alpha} \geq \lambda_i^{\alpha}, -(1 - \xi_{ij}^{\alpha})M, \qquad \forall i \in \mathcal{N}, j \in \mathcal{J}(i), \tag{7.121}$$

$$\Xi_{ij}^{\beta} \leq \Gamma_{ij}^{\beta} M, \qquad \forall i \in \mathcal{N}, j \in \mathcal{J}(i), \tag{7.122}$$

$$\Xi_{ij}^{\beta} \leq \lambda_{ij}^{\beta}, \qquad \forall i \in \mathcal{N}, j \in \mathcal{J}(i), \tag{7.123}$$

$$\Xi_{ij}^{\beta} \geq \lambda_{ij}^{\beta} - (1 - \Gamma_{ij}^{\beta})M, \qquad \forall i \in \mathcal{N}, j \in \mathcal{J}(i), \tag{7.124}$$

$$\lambda, \mu, \phi, \eta, \delta, \Gamma^{\alpha}, \Xi \geq 0, \tag{7.125}$$

$$\varepsilon \text{ unrestricted,} \tag{7.126}$$

$$\xi, \Gamma^{\beta} \in \{0, 1\}. \tag{7.127}$$

7.4.7 Results for an Illustrative Example

To illustrate this analysis, consider the simple network in Fig. 7.3 consisting of seven nodes and seven edges. This is a small network, although the approach can analyze hundreds of nodes and edges with only a few seconds of computation time. For this example, $\Gamma_{tot} \in \{0, 1, \ldots, 14\}$, as there are an α and a β parameter for each of the seven edges. Table 7.4 was created using the above robust approach and solving the 28 optimization programs (a dual and a primal for each level of Γ_{tot}). To recreate Table 7.4 using only the primal formulation (Bertsimas and Sim 2004), the user would solve and search through 6,144 optimization programs; one for each possible combination of Γ_i^{α} and Γ_{ij}^{β} such that their sum is less than Γ_{tot}. For this analysis, it is assumed that $c_i = 1$ for all i and $B = 2$, i.e., at most two nodes can receive defensive investments.

Table 7.4 shows the progression of mission operability, P_1, decreasing as the number of uncertain coefficients, which can result in the worst-case scenario for each value, Γ_{tot}, increases. If $\Gamma_{tot} = 1$, the best investment that BLUE can make will yield a mission operability of 70.5, and this occurs if α_{25} is the parameter having its worst assumed value. If a different parameter had assumed its worst value, the operability could only be higher. The best decisions, which minimize the impact under the possibility that all of the coefficients happen to be their worst values, result in the optimality dropping to as low as 58.4. There is a 13% drop in optimal MOE for $\Gamma_{tot} = 1$ compared with $\Gamma_{tot} = 0$ and a 28% drop for $\Gamma_{tot} = 14$ compared with $\Gamma_{tot} = 0$. It is also noteworthy that the unknowns stop impacting optimality past $\Gamma_{tot} = 5$. This suggests that the best BLUE defensive investment, which hedges against five parameter changes, is unchanged if BLUE wants a more conservative hedge.

Table 7.4 Solutions for the robust mission defense problem for the network in Fig. 7.3 using the new robust approach

Γ_{tot}	Objective value	α changes	β changes
0	81.1		
1	70.5	α_{25}	
2	62.5	α_{25}	β_{12}
3	60.8	α_{25}, α_{26}	β_{12}
4	58.6	α_{25}, α_{26}	β_{12}, β_{25}
5, 6, ..., 14	58.4	$\alpha_{12}, \alpha_{13}, \alpha_{14}$	β_{13}, β_{14}

7.5 Conclusion

This chapter extended the robust optimization techniques in Bertsimas and Sim (2004), given by (7.14)–(7.20), to permit the analysis of uncertain mission networks for which, to be practically useful, the robustness had to be across all constraints and not be merely constraintwise. Rather than making decisions based on nominal dependency parameter assumptions, the derived robust decisions in this chapter will be guarding against some of the assumed parameters deviating from their nominally assumed values. Using the parameter Γ_{tot}, the level of conservatism in investment hedges can be specified from assuming a worst-case possibility to assuming all parameters realizing their nominal values.

This was done with a series of steps in Sect. 7.2. Step 1 implemented the approach in Bertsimas and Sim (2004) and was applicable to a general linear program. Step 2 extended the analysis of a general linear program to make the fidelity of user decisions more nuanced, with a focus on the total number of parameters which might differ from their assumed nominal values rather than the total number of such parameters for each constraint. Step 3 circumvented a technical hurdle, the nonlinearity of the formulation, to permit a faster computational solution. This solution was illustrated with an example in Sect. 7.3.2, which highlighted that even for a small problem with 50 variables and 50 constraints, approximately 80 orders of magnitude of optimization computations could be saved with the approach presented in this chapter. More generally, with m constraints and n variables, the number of required optimizations is reduced from $(n + 1)^m$ to $2mn$. In addition, a measure of network resiliency of potential value emerged from the numerical example on value in its own right.

Section 7.4 specialized the analysis of a general linear program to frame the optimization problem of a defender of a mission network, who must select a level of defensive node investments while hedging against a specified number of parameters capturing dependencies relationships which are not known precisely. Here, for illustrative purposed an FDNA framework to capture dependencies was assumed. The final optimization formulation is (7.104)–(7.127). For completeness, in Sect. 7.4.7 a small robust mission network example is numerically solved.

References

Bertsimas D, Sim M (2004) The price of robustness. Oper Res 52(1):35–53

Ben-Tal A (1985) The entropic penalty approach to stochastic programming. Math Oper Res 10(2):263–279

Ben-Tal A, Nemirovski A (1999) Robust solutions of uncertain linear programs. Oper Res Lett 25(1):1–13

Birge JR, Louveaux F (2011) Introduction to stochastic programming, 2nd edn. Springer, Berlin

Collins JC (2001) Good to great: why some companies make the leap. . . and others don't. HarperBusiness, New York

Feller W (1968) An introduction to probability theory and its applications, vol. 1, 3rd edn. Wiley, New York, p 38

Garvey PR, Pinto CA (2009) Introduction to functional dependency network analysis. In: Proceedings of the second international symposium on engineering systems, Massachusetts Institute of Technology, Cambridge, MA

McCormick GP (1976) Computability of global solutions to factorable nonconvex programs: part I-Convex underestimating problems. Math Program 10(1):147–175

Mulvey JM, Vanderbei RJ, Zenios SA (1995) Robust optimization of large-scale systems. Oper Res 43(2):264–281

Prékopa A (2013) Stochastic programming. In: Mathematics and its applications, vol. 324. Springer Science & Business Media, Berlin

Servi LD, Garvey PR (2017) Deriving global criticality conditions from local dependencies using functional dependency network analysis (FDNA). Syst Eng 20(4):297–306

Soyster AL (1973) Convex programming with set-inclusive constraints and applications to inexact linear programming. Oper Res 21(5):1154–1157

Chapter 8
Robust Mission Network Analysis: The Practice

Once you have mastered a technique, you hardly need look at a recipe again.

Julia Child (2000)

8.1 Introduction

Mission network analysis requires the quantification of the dependencies between physical nodes of a mission. Its representation provides for valuation of the impact of node operability on mission performance to aid preemptive and reactive decision-making. Unfortunately, mission network representations have inherent uncertainty. This uncertainty requires robust valuation methods for informed decisions, but unfortunately such methods are frequently skipped.

This chapter addresses this omission by presenting optimization techniques that hedge against mission network dependency uncertainty and account for the uncertainty costs of defending or attacking a network. This approach produces actionable information for the defender, such as optimal reconfigurations to bolster the network or identifying optimal nodes to harden. For the attacker, it provides equally actionable information such as optimal combinations of nodes to infiltrate and disable. This chapter also presents methods for comparing the robustness and likelihood of optimality of different strategies. It ends with the use of game theory to model how a defender and an attacker may interact by anticipating each other's reactions and starts to explore structural attack deterrence architectures in general and antifragile architectures in particular.

Chapter 7 describes a theoretical approach for hedging against the uncertainty of mission network defense analysis by leveraging and extending the work of Bertsimas and Sim (2004). The approach in Bertsimas and Sim (2004) implicitly assumes that the decisions are continuous ones (so there would be no duality gap created when the dual is computed). This chapter assumes that the decisions are discrete, so the analysis in Chap. 7 is only approximate. Therefore, a new approach is necessary. The new approach has the advantage of being amenable to both defensive

L. Servi, *Mission Dependency Network Analysis*, Risk, Systems and Decisions,
https://doi.org/10.1007/978-3-032-11260-6_8

and offensive mission analysis, as well as the interaction between the defender and the attacker. The approach centers on a new algorithm introduced in Sect. 8.2.2, which is an iterative approach rather than the solution to a Mixed Integer Linear Program (MILP). As such, it might be viewed more as a "cookbook" approach. By its nature, however, it should be viewed more as a guide to approaches taken for such challenges.

The remainder of this chapter is organized as follows: The rest of this section reviews approaches to quantify network dependencies ending with a focus on Functional Dependency Network Analysis (FDNA) (Garvey and Pinto 2009; Servi and Garvey 2017) to illustrate the methods of this chapter. This entails presenting its defining dependency equation and then precisely framing an optimization associated with finding the best defensive or offensive decisions for a network whose dependencies are defined in terms of FDNA. The section ends with a simple illustrative numerical example. Section 8.2 introduces uncertainty of knowledge and demonstrates how to efficiently find the best attack and defense strategies which are robust to this uncertainty. Section 8.3.1 quantifies the properties of the resulting robustness solution. In Sect. 8.3.2, a short excursion is made to demonstrate how one can easily alter the problem formulation to account for possible network reconfigurations to capture redundancy and resilience inherent to the network structure. Finally, Sect. 8.4 uses game theory to examine the case where the defender makes decisions after anticipating the response of the attacker. Section 8.3.2 also introduces the possibility that the system may have antifragile properties (Taleb 2012), which might have the effect that a disruption can paradoxically ultimately strengthen the defender's mission. For example, a disruption makes retiring legacy and modernizing architecture economically inevitable ultimately leaving the system stronger than its pre-attack baseline.[1]

8.1.1 *Quantifying Network Dependencies*

This chapter examines the use of dependency mapping for mission network representations. The goal is to obtain the methodical means of quantifying and reassessing the mission value or the Measure of Effectiveness (MOE). The MOE defines the level at which the system can operate on a relative scale. Various discrete changes, such as removal of nodes or reconfiguration of the network, will raise or lower the expected MOE, allowing the impacts of scenarios to be compared in a systematic way. Figure 8.1 depicts the hierarchical structure of a mission network

[1] For example, the 2007 distributed denial-of-service (DDoS) attacks, Estonia expanded CERT-EE, instituted cross-sector exercises (e.g., Locked Shields) and modernized network practices. These upgrades unlikely, pre-shock, shortened recovery, broadened detection, and reduced mission-failure risk, leaving the system stronger than its pre-attack baseline.

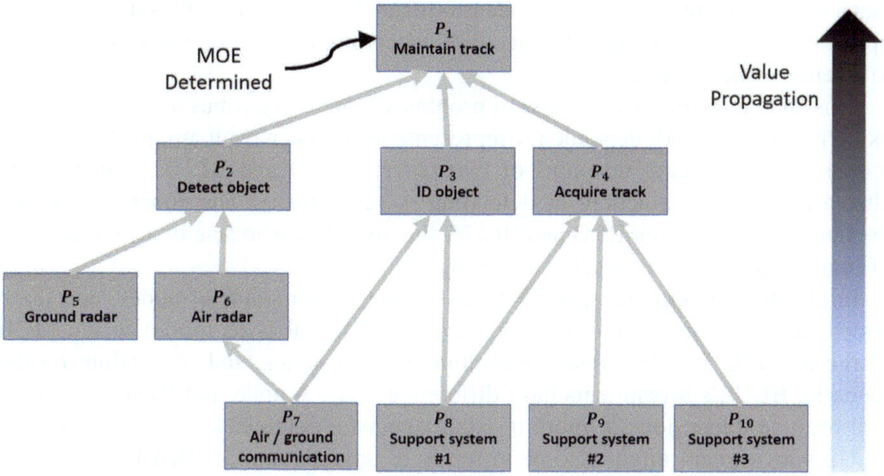

Fig. 8.1 An example of hierarchical dependencies used to represent a system. The value propagates from bottom to top, culminating in the system MOE

and how a value propagates. There are many dependency representations, and the best choice differs by scenario.

As introduced earlier in Chap. 3, one method of mission network valuation and organization is RiskMAP. RiskMAP is used in Crown Jewel Analysis, a process for identifying the cyber assets that are most critical to the success of a mission (Watters et al. 2009). The RiskMAP utility lies with its ease of implementation, but that simplicity precludes more subtle analyses. FDNA is another option (Garvey and Pinto 2009; Servi and Garvey 2017). It is more flexible in nature, giving the user more granular control to define more subtle dependency relationships between systems in the network model. The versatility of FDNA comes from setting two continuous parameters for each edge in the mission network that dictate the dependency relationship between the nodes. System Operational Dependency Analysis (SODA) is a variation of FDNA that adds two more parameters and incorporates a nonlinear component (Guariniello and DeLaurentis 2017). SODA therefore has more granular control and flexibility than FDNA, but parameterization becomes a much more difficult task with four parameters. For simplicity of explanation and interpretation, this chapter will present the algorithms and analytical procedures using FDNA, although the methods presented can be adapted for use with SODA.

FDNA is a dependency representation of a mission network, which quantifies the rolling up of operability across a topology of child-parent dependency relationships. The operability of a parent node is dependent on and determined by the operability of its child nodes. The dependency relationships are characterized by two parameters: α and β as defined in (8.1). The parameter α captures the aggregate impact of all of the dependent children of a node, whereas the parameter β captures the extent that a single dependent child of a node is critical. The mission MOE is defined as the operability of the top node in the network, where higher mission values correspond to a better working system.

This chapter addresses the class of problems centered on robust decision-making using mission network dependency representations. When evaluating networks, we model the actions and reactions of two players: RED and BLUE, who want to alter a network by attacking or defending it, respectively. The equations capture the interactions between RED and BLUE, the impact of who acts first, and the level of impact each player is capable of having.

In all of the interactions, RED attacks some combination of nodes, and these nodes have their operabilities diminished. RED's goal is to minimize the MOE of the network. BLUE invests in combinations of nodes, and, depending on the scenario, BLUE's investments have different effects. A node that BLUE invests in will become impervious or more resistant against a potential attack. The effect of BLUE's investment might also be to repair or upgrade a node. BLUE's goal is to maximize the MOE of the network.

Table 8.1 lists the notation needed to make the analysis precise.

8.1.2 The Master FDNA Equation

Equation (8.1) is the FDNA master equation that characterizes a parent node's operability in terms of the operability of its child nodes (Garvey and Pinto 2009; Servi and Garvey 2017). In this and subsequent equations in this chapter, i denotes the parent node that depends on support from child node j. One might interpret (8.1) as stating that the operability of parent node i equals the minimum of two different dependency influences: (i) a given constant, β_{ij}, above the operability of each of its child nodes, $j \in \mathcal{J}(i)$, and (ii) a weighted average of the operability of all of the child nodes according to α_{ij}. The β parameter captures the possible fragility of node i where a single dependent child is critical. The weighted average portion captures the relative influence of each of the dependent children nodes. In the special case where node i is a leaf node, its operability level is based on its baseline operational level and the actions of RED and BLUE.

$$P_i = \min \left[100 - \frac{1}{|\mathcal{J}(i)|} \sum_{j \in \mathcal{J}(i)} \left(100 - P_j\right) \alpha_{ij}, \min_{j \in \mathcal{J}(i)} \left[P_j + \beta_{ij}\right] \right], \tag{8.1}$$

where $\mathcal{J}(i)$ characterizes the set of nodes which impact the MOE of node i.

Table 8.1 Notation for optimization program formulation

$\vec{w}$	The vector of BLUE's *investments* where $w_j = 1$ if BLUE invests in node j
$\vec{z}$	The vector of RED's *attacks* where $z_j = 1$ if RED attacks node j
c_j^w	The direct costs of BLUE's investment in node j
c_j^z	The direct cost for RED to attack node j
B	The budget by which RED's attacks or BLUE's investments are constrained
B_ℓ	The lower bound on B
B_h	The upper bound on B
ψ^w	The propensity for BLUE to invest in any given node (in units of MOE/$)
ψ^z	The propensity for RED to attack any given node (in units of MOE/$)
γ_i	The base operability of node i in the absence of any alteration
P_j	The measure of effectiveness (MOE) or operations level of node j
$\mathcal{N}$	The set of all nodes
$\mathcal{L}$	The set of leaf nodes
n	The number of non-leaf nodes: $\lvert\mathcal{L}^c\rvert$
$\mathcal{J}(i)$	Characterizes the set of nodes which impact the MOE of node i. For leaf nodes $\mathcal{J}(i) = \{\}$
α_{ij}	The weighted average dependency parameter capturing dependency of parent node i on child node j
β_{ij}	The scalar difference dependency parameter capturing dependency of parent node i on child node j
h_i	The maximum value of node i, given investment by BLUE, as a proportion of the baseline operability
	The default value is $h_i = 1\ \forall i$ but $h_i > 1$ could be used to model network upgrades
ℓ_i	The lower bound the operability of node i as a proportion of the baseline operability
	If node i is not attacked: $\ell_i = 1$; else $0 \leq \ell_i < 1$
	The latter captures modeling attacks that may not fully disable nodes
I_i^1, I_{ij}^2	Binary variables used in attack formulation to identify which part of the FDNA master equation is binding
M	A large positive number used in conjunction with the I_i^1 and I_{ij}^2 variables

8.1.3 *Defender Formulation*

The defender formulation starts with the notation in Table 8.1 and its objective: The objective function (8.2) maximizes BLUE's utility by selecting the level of defensive investment for each node, w_i, as well as BLUE's total costs, B, in order to capture the trade-offs between B and the MOE of the mission network (which, by convention, is the operability value of node 1, P_1). The choice of ψ^w dictates the relative value of the cost of investing in a node. As ψ^w increases, BLUE's recommended defensive strategy will entail investing less. In some cases, the budget for BLUE's total costs is fixed. In this case ψ^w is irrelevant, and one should set $B_\ell = B_h$.

More formally, the optimization in (8.2)–(8.8) captures these trade-offs of protecting nodes $\{i : \ell_i < 1\}$ that might be attacked by RED:

$$\max_{w,B,P} \left[P_1 - \psi^w B \right] \tag{8.2}$$

subject to

$$P_i \leq P_j + \beta_{ij}, \qquad \forall i \in \mathcal{L}^C, j \in \mathcal{J}(i), \tag{8.3}$$

$$P_i \leq 100 - \frac{1}{|\mathcal{J}(i)|} \sum_{j \in \mathcal{J}(i)} \alpha_{ij}(100 - P_j), \qquad \forall i \in \mathcal{L}^C, \tag{8.4}$$

$$P_i \leq \ell_i \gamma_i + (h_i - \ell_i) w_i \gamma_i, \qquad \forall i \in \mathcal{L}, \tag{8.5}$$

$$\sum_i c_i^w w_i \leq B, \tag{8.6}$$

$$B_\ell \leq B \leq B_h, \tag{8.7}$$

$$w_i \in \{0, 1\}, \qquad \forall i. \tag{8.8}$$

Equations (8.3) and (8.4) enforce the structure of FDNA valuation described in (8.1). Equation (8.6) ensures that the sum of the defensive investments made does not exceed the budget B chosen by the optimization. Equation (8.7) allows the user to choose hard bounds for B overriding the optimal choice based on the preference ψ^w. Alternatively, the user could set $B_\ell = 0$ and $B_h = \sum_i c_i^w$ to ensure that all decision-making is completely driven by the value of ψ^w and not any artificial constraints on B.

Equation (8.5) is summarized in Table 8.2 and contains the heart of the optimization and is most easily understood by considering the four possibilities shown in Table 8.2. If leaf node i is attacked, i.e., $0 \leq \ell_i < 1$, it ensures that it will have value $\ell_i \gamma_i$ if node i is not invested in, i.e., $w_i = 0$, and is $h_i \gamma_i$ if node i is invested in, i.e., $w_i = 1$. The parameter h_i can be set to any positive value. If $h_i > 1$, the node will have a value greater than its nominal value due to a possible upgrade from BLUE's investment. If leaf node i is not attacked, i.e., $\ell_i = 1$, then it will have value γ_i if not defended, i.e., $w_i = 0$ and, again, $h_i \gamma_i$ if defended, i.e., $w_i = 1$.

Table 8.2 Impact of operability of node i depending on whether or not RED attacks it and whether or not BLUE defends it

	Defended ($w_i = 1$)	Not defended ($w_i = 0$)
Attacked ($0 \leq \ell_i < 1$)	$h_i \gamma_i$	$\ell_i \gamma_i$
Not attacked ($\ell_i = 1$)	$h_i \gamma_i$	γ_i

8.1.4 Attacker Formulation

RED's goal is to minimize the mission value while accounting for the varying costs for different attacks. Although the user can specify a budget by setting $B_\ell = B_h$, the objective function (8.9) more generally maximizes the trade-off impact on the MOE with the total cost of a set of attacks. The choice of ψ^z dictates the relative value of attacking a node. If the decrease in P_1 exceeds the increase in $\psi^z B$, then RED attacks the node.

$$\min_{z,B,I^1,I^2,P} \left[P_1 + \psi^z B \right] \tag{8.9}$$

subject to

$$P_i \geq P_j + \beta_{ij} - (1 - I^2_{ij})M, \qquad \forall i \in \mathcal{L}^{\mathsf{C}}, j \in \mathcal{J}(i), \tag{8.10}$$

$$P_i \geq 100 - \frac{1}{|\mathcal{J}(i)|} \sum_{j \in \mathcal{J}(i)} \alpha_{ij}(100 - P_j) - (1 - I^1_i)M, \qquad \forall i \in \mathcal{L}^{\mathsf{C}}, \tag{8.11}$$

$$I^1_i + \sum_{j \in \mathcal{J}(i)} I^2_{ij} = 1, \qquad \forall i \in \mathcal{L}^{\mathsf{C}}, \tag{8.12}$$

$$P_i \geq \gamma_i - z_i \gamma_i (1 - \ell_i), \qquad \forall i \in \mathcal{L}, \tag{8.13}$$

$$\sum_i c^z_i z_i \leq B, \tag{8.14}$$

$$B_\ell \leq B \leq B_h, \tag{8.15}$$

$$z, I \in \{0, 1\}. \tag{8.16}$$

Equations (8.10), (8.11), and (8.12) enforce the structure of the FDNA valuation (8.1). The formulation includes the big M (a sufficiently large positive constant for FDNA is $M = 100$) and binary I variables to indicate the binding constraint of (8.1), because the optimization problem is a minimization and the FDNA equations contain "Min" arguments. For each value of i, there is one I^1_i variable, and there are $|\mathcal{J}(i)|$ I^2_{ij} variables. If $I^1_i = 1$, the α_i constraint in (8.1) is the binding constraint. If $I^2_{ij} = 1$, the β_{ij} constraint is the binding constraint. Equation (8.14) captures the total attack costs B, which is included in the objective function. Equation (8.15) allows the user to choose hard bounds on the total cost of attacks. Alternatively, the user can set $B_\ell = 0$ and $B_h = \sum_i c^z_i$ to ensure that all decision-making is driven by the value of ψ^z.

Equation (8.13) is summarized in Table 8.2 and contains the heart of the optimization and is most easily understood by considering the four possibilities shown in Table 8.2. If leaf node i is attacked, i.e., $z_i = 1$, then from (8.13) $P_i = \ell_i \gamma_i$.

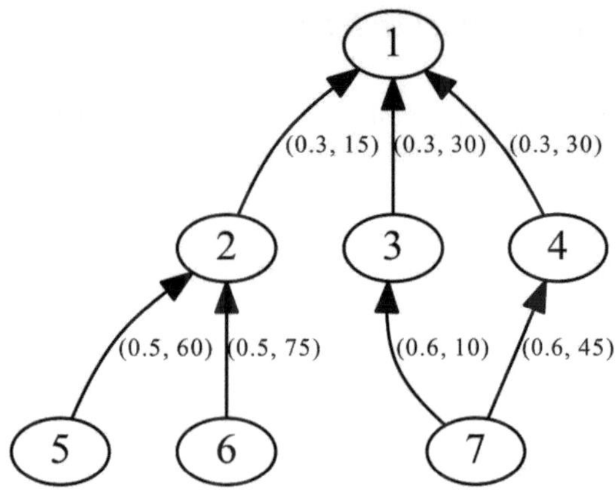

Fig. 8.2 A small network used to demonstrate mission network analysis. The edge labels are the corresponding (α, β) dependency parameters. Note that the arrows flow from child node to parent node corresponding to the direction of value propagation through the network. The baseline operating levels for leaf nodes 5, 6, and 7 are assumed to be 45, 25, and 80, respectively

The value of ℓ_i will be specified before the optimization is computed: If node i is not defended, ℓ_i will be set to h_i. If node i is not attacked, i.e., $z_i = 0$, then from (8.13) $P_i = \gamma_i$. To be consistent with the defender formulation, the value of γ_i will be set to $h_i\gamma_i$ if node i is not defended.

8.1.5 A Small Example of Solving the Attack Formulation with no Hedging

Consider the network in Fig. 8.2. For this example, we focus on RED's perspective by evaluating attacks on the network. For simplicity of exposition, we assume $B_\ell = B_h$, which fixes B and makes ψ^w irrelevant to the objective function. Also for simplicity, we set the cost of attacking each node equal to each other by presuming that for each node, $c_i^z = 1$. We also assume an attacked leaf node will send its operability to 0, i.e., $\ell_i = 0$ for all i. In this section, it is assumed that there is no uncertainty in any of the formulation's parameters. As such, there is no need to seek an optimal decision that hedges against what is known to be unknown.

Table 8.3 shows the node operability levels under the four different budget levels for RED. The gray boxes indicate the optimal choice of node or nodes that RED should attack under each scenario. The entries in Table 8.3 are the operability of each node after the attack according to the FDNA equation (8.1), assuming that the nodes corresponding to the gray boxes are attacked.

Note how, as expected, the top row (corresponding to the MOE of the network) is nonincreasing as the RED's budget increases. When $B = 0$, corresponding to not attacking any nodes, the network operates with a mission value of 82.5. If RED is allowed to attack one node, RED should choose P_7 and drop the mission value to 70. If RED is allowed to attack two nodes, RED will choose nodes {5, 6} and drop the mission value to 65. Even if RED is allowed to attack all three leaf nodes, the mission value will remain unchanged at 65. One might note, parenthetically, that

Table 8.3 Optimal attack combinations to minimize the mission value at each level of B for the network in Fig. 8.2 with no uncertainty in dependency parameters

	$B = 0$	$B = 1$	$B = 2$	$B = 3$
P_1	82.5	70	65	65
P_2	67.5	67.5	50	50
P_3	88	40	88	88
P_4	88	40	88	88
P_5	45	45	0	0
P_6	25	25	0	0
P_7	80	0	80	0

the best strategy for RED with a budget of 2 is not myopic, i.e., RED should not choose the best node, node 7, to attack with a budget of 1 along with another node, but rather attack two different nodes, 5 and 6.

If the budget is not restricted to a single value using B_ℓ and B_h, the optimization will find the best budget if the user specifies the trade-off between the relative importance of impact on the mission, P_1, and the budget, B, using the parameter ψ^z. As such, ψ^z will drive the best value of B. If $\psi^z > 12.5$, the best decision is to set $B = 0$, which results in MOE = 82.5. For $\psi^z \in (5, 12.5]$, the best budget is $B = 1$, which results in MOE = 70. Finally, if $\psi^z \leq 5$, then $B = 2$ and MOE = 65.

8.2 Finding the Robust Optimal Solution

Robust optimization of mission networks is important because estimates of the dependency parameters in dependency representations such as FDNA are uncertain; such uncertainty can propagate throughout the network, sometimes causing a disproportionate impact. To make informed decisions, one should hedge against this uncertainty. In short, one should sacrifice some impact assuming no uncertainty in return for more robustness in the impact of decisions. It is better to make a decision robust against much worse outcomes, which may come to fruition within the levels of uncertainty present in the valuation. The Robust Optimization Solution (ROS) refers to the best robust solution, and, for a baseline, the Nominal Optimal Solution (NOS) refers to the optimal solution when we assume no uncertainty.

One approach to handling robustness is to use stochastic programming to optimize the uncertain mission network parameters such that the solution is feasible with some probability (Birge and Louveaux 2011; Prékopa 2013). Stochastic programming requires probability distributions for the uncertainty, which is often too restrictive for mission networks. Additionally, stochastic programming allows constraints to be violated, which can be unacceptable in pass/fail scenarios. Ben-Tal and Nemirovski's approach is to offset this with entropic penalties for solutions, which are not "feasible in the mean," but this still allows for some degree of constraint violation (Ben-Tal 1985). Bertsimas and Sim made a large leap toward solving this type of problem by introducing a parameter for the level of conservatism for each constraint in a linear program (Bertsimas and Sim 2004). The conservatism parameter, Γ_i, is the number of coefficients in constraint i (corresponding, in the mission network context, to dependency influences on parent node i) that are allowed to vary within a specified interval. Their method produces a solution that is feasible in the worst-case scenario for that level of conservatism.

The idea of using a tuning parameter for the balance between conservatism and robustness (Bertsimas and Sim 2004) will be used in this chapter. Unfortunately the approach of Bertsimas and Sim (2004) is insufficient specifically for mission network analysis because we need to control the total number of parameters across *all* constraints. In particular, since for mission network analysis, half of the constraints contain only one parameter with uncertainty, and the original approach in Bertsimas and Sim (2004) is not helpful.

In this chapter, we use the concept of robustness created by Bertsimas and Sim but expand the definition. When we refer to the ROS, we are referring to the optimal solution given that Γ_{tot} parameters (throughout the entire program) are within their assumed uncertainty range to create the worst-case scenario possible and the other parameters are assumed to be at their nominal values. The Γ_{tot} parameter is an integer representation of the conservativeness of the scenario, where large values correspond to very conservative scenarios. It might be useful to define uncertainty in each parameter separately: $\Gamma_{tot} = \Gamma_\alpha + \Gamma_\beta$, where Γ_α and Γ_β are the total number of α and β parameters allowed to vary. The robust formulation for both RED (the attacker) and BLUE (the defender) follows the same procedure with minor alterations. For simplicity of exposition, in this section we present the procedure from RED's perspective.

The algorithmic approach of Bertsimas and Sim (2004), which was generalized in Chap. 7, involved taking the dual of part of the problem formulation. This approach implicitly assumed that the decisions were continuous ones (so there would be no duality gap created when the dual is computed). This chapter assumes that the decisions are discrete, so the analysis of Chap. 7 and Bertsimas and Sim (2004) is not applicable. Therefore, a new approach is needed that has the advantage of being amenable to both defensive and offensive mission analyses, as well as the interaction between the defensive and offensive actors. The approach centers on an iterative approach rather than the solution to a single Mixed Integer Linear Program (MILP) and is presented in Sect. 8.2.2.

8.2.1 Uncertainty

Table 8.4 contains notations used to formulate the robust program that incorporates uncertainty. The uncertainty of interest is due to a lack of perfect knowledge of a system rather than random uncertainty due to the stochasticity of an event. As such, it is captured by a range of dependency parameter values without assigning probabilities within that range.

Table 8.4 Notation for robust optimization program formulation

Γ_α	The integer quantity of α parameters allowed to vary within their uncertainty ranges
Γ_β	The integer quantity of β parameters allowed to vary within their uncertainty ranges
Γ_{tot}	The integer quantity dependency parameters allowed to vary within their uncertainty ranges $\Gamma_{tot} = \Gamma_\alpha + \Gamma_\beta$.
a_{ij}	A binary indicator signifying that the α_{ij} parameter is incorrect and should be augmented by $\widetilde{\alpha}_{ij}$
b_{ij}	A binary indicator signifying that the β_{ij} parameter is incorrect and should be augmented by $\widetilde{\beta}_{ij}$
$\widetilde{\alpha}_{ij}$	The extent of uncertainty in dependency parameter α_{ij}
$\widetilde{\beta}_{ij}$	The extent of uncertainty in dependency parameter β_{ij}
c_i	The cost to the attacker of attacking node i
q_{ij}	A linearizing variable that equates to the quantity of $P_j a_{ij}$
$\mathcal{Z}$	The set of unique optimal attack vectors at each step of the iterative algorithm = $\{\vec{z}^1, \vec{z}^2, \ldots, \}$

The following is the formal formulation for finding the attacker's (i.e., RED's) ROS. The attacker must decide which nodes to degrade given the knowledge of all of the dependency variables, while hedging against the knowledge that the worse choice, from the attacker's point of view, of the number, Γ_α, of α parameters, and the number, Γ_β, of β parameters will deviate from their nominal values. The formulation is an extension of (8.9)–(8.16).

$$\mathcal{R}(\vec{z}, \vec{a}, \vec{b}, \mathcal{Z}) = \max_{a,b} \left[\min_{z,B,I^1,I^2,P} \left(P_1 + \psi^z B \right) \right] \tag{8.17}$$

subject to

$$P_i \geq P_j + \beta_{ij} + \widetilde{\beta}_{ij} b_{ij} - (1 - I_{ij}^2) M, \qquad \forall i \in \mathcal{L}^{\mathrm{c}},\ j \in \mathcal{J}(i), \tag{8.18}$$

$$P_i \geq 100 - \frac{1}{|\mathcal{J}(i)|} \sum_{j \in \mathcal{J}(i)} \left(\left(\alpha_{ij} + \widetilde{\alpha}_{ij} a_{ij} \right) \left(100 - P_j \right) \right) - \left(1 - I_i^1 \right) M,$$
$$\forall i \in \mathcal{L}^{\mathrm{c}}, \tag{8.19}$$

$$I_i^1 + \sum_n I_{ij}^2 = 1, \qquad \forall i \in \mathcal{L}^{\mathrm{c}}, \tag{8.20}$$

$$P_i \geq \gamma_i - z_i \gamma_i (1 - \ell_i), \qquad \forall i \in \mathcal{L}, \tag{8.21}$$

$$\sum_{i \notin \{\hat{i}: z^k_{\hat{i}}=1, \vec{z}^k \in \mathcal{Z}\}} z_i \geq 1, \qquad \forall k, \tag{8.22}$$

$$\sum_i c_i z_i \leq B, \tag{8.23}$$

$$B_\ell \leq B \leq B_h, \tag{8.24}$$

$$\sum_{i \in \mathcal{L}^c, j \in \mathcal{J}(i)} a_{ij} \leq \Gamma_\alpha, \tag{8.25}$$

$$\sum_{i \in \mathcal{L}^c, j \in \mathcal{J}(i)} b_{ij} \leq \Gamma_\beta, \tag{8.26}$$

$$z, I, a, b \in \{0, 1\}. \tag{8.27}$$

Suppose that $\vec{z}^k = (z^k_1, z^k_2, \ldots,)^T$ is an element of the set of vectors, $\mathcal{Z}$, where $\vec{z}^k$ was an interim attack vector previously identified. In (8.17) and (8.27), z, I, a, b, and P refer to the vector or matrix of individual components. The optimization in (8.17) has a maximization of a minimization problem that ensures the worst-case uncertainty scenario for the attacker given their decision. Hence, the solution guides the attacker how to more robustly attack the network.

The internal minimization program in (8.17) selects which nodes to attack, i.e., it selects $\vec{z}$, to degrade and minimize the operability of the network mission P_1, and the budget B used. As in (8.9)–(8.16), to do so, the budget B has to also be selected to optimize the trade-off between costs and mission impact. Also, the program has to determine which part of the FDNA formula (8.1) is binding, i.e., it selects the values of I^1 and I^2. In addition, the program determines the operability of all the nodes, P

Equations (8.18) and (8.19) combine to account for uncertainty in the β parameters with β_{ij}, from (8.10), being replaced with $\beta_{ij} + \tilde{\beta}_{ij} b_{ij}$. They ensure that no more than Γ_β parameters change, but that the effect of the changes maximizes the objective function. Similarly, (8.19) accounts for uncertainty in the α parameters with α_{ij}, from (8.11), first being replaced with $\alpha_{ij} + \tilde{\alpha}_{ij} a_{ij}$ and then using the argument in the above paragraph.

Equations (8.20), (8.21), (8.23), (8.24), and (8.27) correspond to constraints (8.12), (8.13), (8.14), (8.15), and (8.16), respectively. Constraint (8.22) is a new constraint. As will be described in Sect. 8.2.2, by definition, $\mathcal{Z}$ is the set of all interim attack vector solutions found in the previous iterations of algorithm, i.e., $\mathcal{Z} = \{\vec{z}^k : \text{for all k}\}$. Initially, $\mathcal{Z}$ is set to the empty set, and the optimization is unconstrained with respect to other iterations of the algorithm. Equation (8.22) ensures that, at the next iteration the algorithm, $\vec{z} = (z_1, z_2, \ldots,)$ will have at least one value of z_i, which is different from each previous iteration optimal attack vector. Equation (8.22) is included to ensure that the algorithm described in Sect. 8.2.2 cannot get stuck in an endless cycle.

In principle (8.17)–(8.27) could form the basis for an iterative algorithm to solve the challenge of finding an algorithm which can deal with the uncertainty. However, unfortunately, (8.19) contains a nonlinearly, the $a_{ij}P_j$ term, which would unnecessarily slow down the computation time to solve (8.17)–(8.27). This complication can be overcome by introducing a new variable, q_{ij}, alone with four new constraints,

$$q_{ij} \leq 100a_{ij}, \qquad \forall i \in \mathcal{L}^{C}, j \in \mathcal{J}(i), \tag{8.28}$$

$$q_{ij} \leq P_j, \qquad \forall i \in \mathcal{L}^{C}, j \in \mathcal{J}(i), \tag{8.29}$$

$$q_{ij} \geq P_j - 100\left(1 - a_{ij}\right), \qquad \forall i \in \mathcal{L}^{C}, j \in \mathcal{J}(i), \tag{8.30}$$

$$q_{ij} \geq 0, \qquad \forall i \in \mathcal{L}^{C}, j \in \mathcal{J}(i), \tag{8.31}$$

which, as will be argued below, along with (8.17)–(8.27) force $q_{ij} = a_{ij}P_j$. Once this is established, (8.19) can be replaced by

$$P_i \geq 100 - \frac{1}{|\mathcal{J}(i)|} \sum_{j \in \mathcal{J}(i)} \left(\alpha_{ij}\left(100 - P_j\right) - \tilde{\alpha}_{ij}\left(100a_{ij} - q_{ij}\right)\right) - \left(1 - I_i^1\right) M,$$

$$\forall i \in \mathcal{L}^{C} \tag{8.32}$$

and thereby eliminate the nonlinearity.

The argument that $q_{ij} = a_{ij}P_j$ is as follows and is based on McCormick (1976). The minimization will be achieved when the right-hand side of (8.32) is as small as possible, which in turn occurs when q_{ij} is as small as possible. From (8.30) and (8.31), $q_{ij} \geq max(P_j - 100(1 - a_{ij}), 0)$, which one can verify equals $a_{ij}P_j$ regardless of whether a_{ij} is 0 or 1. The outer maximization program in (8.17) determines where uncertainty is realized within the network in order to most deter the attacker's goal of minimizing the mission objective, i.e., it selects $\vec{a}$ and $\vec{b}$. The maximin is achieved where the feasible region of the minimization is as small as possible. Therefore, the outer maximization will select $\vec{a}$ so that the right-hand side of (8.32) is as large as possible, which in turn occurs when q_{ij} is as large as possible. From (8.28) and (8.29), $q_{ij} \leq \min(100a_{ij}, P_j)$, which again equals $a_{ij}P_j$ regardless of whether a_{ij} is 0 or 1. Since we showed that $q_{ij} \geq a_{ij}P_j$ and $q_{ij} \leq a_{ij}P_j$, it follows that $q_{ij} = a_{ij}P_j$.

With a slight oversimplification of notation by suppressing some variables to illuminate clarity, define $\mathcal{R}(\vec{z}, \vec{a}, \vec{b}, \mathcal{Z})$ to be the value of the objective in (8.17), subject to the constraints of (8.18)–(8.31), with (8.19) replaced by (8.32). Also view $\max_{\vec{a},\vec{b}} \min_{\vec{z}} \mathcal{R}(\vec{z}, \vec{a}, \vec{b}, \{\})$ as the optimal solution. The optimal solution is the best decision for the attacker given the dependency relationships of the mission network, *as well as* given that Γ_{tot} parameters change (throughout the entire program) within their uncertainty range to create the worst-case scenario possible. We refer to this as the Robust Optimization Solution (ROS). As will be explained below, the set of vectors, $\mathcal{Z}$, will be used to find the ROS.

One possible method for solving these types of maximin problems is to take the dual of the inner problem so that the program becomes a maximization of a maximization. That was the approach taken in Chap. 7. However, due to the integrality gap caused by the integer decision variables, we cannot use the dual, so this method will fail. Alternatively, one might seek to solve the program iteratively, i.e., solve the inner minimization and use that solution to seed the outer maximization, which in turn seeds the inner minimization, etc. Unfortunately, such an approach typically fails as it frequently results in cycles. The cycles, however, can be broken with a simple alteration of the algorithm, which constrains the minimization to never find the same solution twice—this is the role of (8.22). The algorithm tracks the number of iterations, as well as the lower and upper bounds of best solutions found to form a basis for a stopping rule. No stopping rule is recommended in this chapter as it depends on the nuances of the mission of interest. In theory, this algorithm may result in an exhaustive exploration of all values of $\vec{z}$, $\vec{a}$, and $\vec{b}$. In practice, the algorithm finds an excellent solution quickly. This algorithm, the Mission Network Pivot Algorithm, is formally presented in Sect. 8.2.2.

8.2.2 *The Mission Network Pivot Algorithm*

The Mission Network Pivot Algorithm seeks to pivot through the solution space and find the attack combination that solves $\max_{\vec{a},\vec{b}} \min_{\vec{z}} \mathcal{R}(\vec{z}, \vec{a}, \vec{b}, \mathcal{Z})$. The first four lines in Algorithm 9 initialize variables used in the robust program. The bulk of the algorithm then contains two main steps. Line 6 finds $\vec{z}$ that minimizes the program, and line 7 updates the lower bound based on the optimal $\vec{z}$. Line 11 uses this solution to find $\vec{a}$ and $\vec{b}$ that maximize the program before line 12 updates the upper bound based on the optimal $\vec{a}$ and $\vec{b}$.

Lines 6 and 7 compute the lower bound encountered so far. Lines 8 and 9 of the Mission Network Pivot Algorithm ensure unique solutions for each iteration of the process. Line 8 checks to see if the current attack vector $\vec{z}^i$ is an element of the set of optimal attack vectors found in the previous steps of the algorithm, $\mathcal{Z}$. If so, the algorithm reoptimizes using $\mathcal{Z}$, so constraint 8.22 will ensure that a new attack vector solution will be identified. Each additional element of $\mathcal{Z}$ further shapes the solution space and allows the algorithm to break out of any cycles we may otherwise enter.

There are a number of reasonable options for the $stop()$ function, and the best choice may change based on the network. One possible stopping criterion is to stop if the reoptimization in line 9 becomes infeasible because all possible attack vectors have been examined. This can be used in a small network like the one in Fig. 8.2. For larger networks, that stopping criterion is too computationally demanding. Other possibility for larger networks is to iterate a set number of times based on the size of the network. Another possible option is to only iterate through the algorithm a certain number of times past the most recent improvement in objective function. That is the stopping criterion used for the analysis presented in this chapter.

Algorithm 9 Mission network pivot algorithm

$k = UB = 0$
$LB = 100$
$\vec{a}^k = \vec{b}^k = \vec{z}^k = \vec{0}$ for all k;
$\mathcal{Z} = \{\}$
while not stop() **do**
  $\vec{z}^k = \arg\min_{\vec{z}} \mathcal{R}(\vec{z}, \vec{a}^k, \vec{b}^k, \{\})$
  $LB = \max\left(\text{LB}, \mathcal{R}(\vec{z}^k, \vec{a}^k, \vec{b}^k, \{\})\right)$
  if $\vec{z}^k \in \mathcal{Z}$ **then**
    $\vec{z}^k = \arg\min_{\vec{z}} \mathcal{R}(\vec{z}, \vec{a}^k, \vec{b}^k, \mathcal{Z})$
  end if
  $\mathcal{Z} = \mathcal{Z} \cup \vec{z}^k$
  $\{\vec{a}^k, \vec{b}^k\} = \arg\max_{\vec{a},\vec{b}} \mathcal{R}(\vec{z}^k, \vec{a}, \vec{b}, \{\})$
  $UB = \min\left(\text{UB}, \mathcal{R}(\vec{z}^k, \vec{a}^k, \vec{b}^k, \{\})\right)$
  $k \leftarrow k + 1$
end while
return $\vec{z}^k, a^k, b^k$

8.2.3 An Illustrative Example

This algorithm was implemented with $\Gamma_\alpha = 2, \Gamma_\beta = 1, c_i = 1$ for all $i, \tilde{\alpha}_{ij} = 0.1\alpha_{ij}$ for all i and j, $\tilde{\beta}_{ij} = 0.1\beta_{ij}$ for all i and j. The topology of the mission network and the nominal values of the α's and β's are given in Fig. 8.2, and $B_\ell = B_h = 2$. For these parameters, the attacker may attack $B_\ell = B_h = 2$ nodes, and the values of the α's and β's may have an error as large as 10% of their values. For this analysis it is viewed as overly conservative to assume that *all* of the α's and β's are in error by 10 percent of their nominal values; instead, at most $\Gamma_\alpha = 2$ of the α's and at most $\Gamma_\beta = 1$ of the β's are in error by 10%. As to which two α's and which β are in error, the answer is that the optimization will identify the ones that most counter the attacker's choices of nodes to attack.

The output sequence of the application of the Mission Network Pivot Algorithm, given the above data description, is found in Table 8.5. Each row corresponds to a new iteration of the algorithm, with the iteration number listed in the first column of the table. The fourth column indicates the best two nodes for the attacker to attack. The second and third columns correspond to which two α values and single β values being in error are most harmful to the attacker's goal of minimizing the mission's MOE. The fifth and sixth columns indicate the algorithm's lower and upper bounds of the best the attacker can achieve. The algorithm stops after three iterations as the upper and lower bounds of the MOE are equal to 75. This is the Robust Optimization Solution (ROS).

In the absence of uncertainty, from Table 8.3, the best action for the attacker is to attack $B = 2$ number of nodes: nodes 5 and 6. In this case, according to Table 8.3, the attacker would expect the impact to be lowering the MOE to 65. However, in the face of uncertainty, that would be incorrect. As can be seen from the first row

Table 8.5 The output generated by Algorithm 9 for the network in Fig. 8.2 given that $\Gamma_\alpha = 2$, $\Gamma_\beta = 1$, and $B = 2$

Algorithm step	α changes	β changes	Attacked nodes	Lower bound	Upper bound
1	$\{\alpha_{25}, \alpha_{26}\}$	$\{\beta_{25}\}$	$\{5, 6\}$	65	87.5
2	$\{\alpha_{25}, \alpha_{37}\}$	$\{\beta_{14}\}$	$\{6, 7\}$	70	85
3	$\{\alpha_{25}, \alpha_{38}\}$	$\{\beta_{15}\}$	$\{5, 7\}$	75	75

of Table 8.5, the attacker might only decrease the MOE to 87.5. This upper bound, in column 6, is the known guaranteed achievable MOE after the first step if α_{25}, α_{26}, and β_{25} were misestimated by 10%. From Table 8.3, the attacker would attack the wrong nodes, 5 and 6, instead of the ROS solution: nodes 5 and 7. As shown in the third row of Table 8.5, with the optimal attack decisions, the MOE could have been guaranteed to be decreased to 75. This is the power of this robust analysis: The solution hedges against the uncertainty of the dependency parameter values to achieve a better guaranteed solution.

8.3 Properties of the Solutions

8.3.1 Randomly Generated Scenario Analysis

In Sect. 8.2, the Robust Optimization Solution was examined. When applied to the attacker problem, this entailed determining the solution, in (8.17), to the optimization program

$$\max_{\vec{a},\vec{b}} \min_{\vec{z}} \mathcal{R}(\vec{z}, \vec{a}, \vec{b}, \{\}).$$

This section examines a complementary approach to provide a different perspective and determine the solutions stochastically. This may suggest an alternative approach to approximating the ROS solution and provide some qualitative insight into the nature of the solutions of this class of problems. Specifically, this section examines the optimal decisions for the attacker given no uncertainty, i.e., the NOS solution, but for thousands of simulated trials where the values of the α's and β's are altered according to the input parameters of the ROS formulation assuming a uniform probability distribution for $\vec{a} \in [\vec{\alpha} - \vec{\tilde{\alpha}}, \vec{\alpha} + \vec{\tilde{\alpha}}]$ and $\vec{b} \in [\vec{\beta} - \vec{\tilde{\beta}}, \vec{\beta} + \vec{\tilde{\beta}}]$. The striking result is that the ROS solution is among the most common optimal decisions for the attacker.

Specifically, we sample $\vec{a}$ and $\vec{b}$ randomly using Monte Carlo-inspired methods and compare them with the optimal attack vector found by the Mission Network Pivot Algorithm described in Sect. 8.2.2. For this, we define $\vec{a}(\omega)$ and $\vec{b}(\omega)$ as randomly generated samples of the dependency uncertainty and

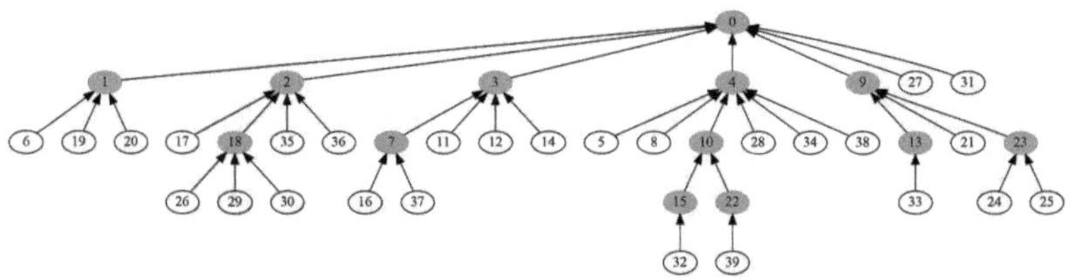

Fig. 8.3 Randomized network with 40 nodes and 39 edges. The edges flow from child nodes to parent nodes. White nodes are leaf nodes, which are commonly the only nodes modeled as subject to attacks

$$\vec{z}(\omega) = \arg\min_{\vec{z}} \mathcal{R}(\vec{z}, \vec{a}(\omega), \vec{b}(\omega), \{\})$$

as the optimal attack vector for the random uncertainty ω and note the most frequent attack vectors found.

We explore the multiplicity of the optimal solutions given a random configuration of uncertainty in the network. However, if the network is sufficiently large, we cannot enumerate all the combinations of dependency parameters that can change. For example, for the network mission of Fig. 8.3, we have $\Gamma_\alpha = \Gamma_\beta = 20$, and the network has 39 edges, i.e., $\binom{39}{20}^2 > 10^{21}$ combinations available. Therefore, we use random sampling to estimate the metrics and probabilities that characterize the solutions and observe the comparative likelihoods of the solution being the Robust Optimization Solution (ROS) found using the Mission Network Pivot Algorithm or the Nominal Optimal Solution (NOS).

To estimate the frequency of optimality for each of the solutions, we sample $\Gamma_\alpha = \Gamma_\beta = 20$ dependency parameters to be randomly altered for a given run and optimize against those with a fixed budget $B = 3$. Each dependency parameter has an equal chance of being selected. Figure 8.4 displays the histogram of the optimal attack vectors using 10,000 independent samples, which empirically estimates $\text{PR}_\omega[\vec{z}(\omega) = \vec{z}_i]$ for each solution $\vec{z}_i$. It might be surprising that only four distinct solutions were found. They include the ROS, $\{6, 21, 26\}$, and the NOS, $\{26, 27, 31\}$. Figure 8.4 indicates that, empirically, the $\text{PR}_\omega[\vec{z}(\omega) = \vec{z}_{ROS}]$ is slightly higher than $\text{PR}_\omega[\vec{z}(\omega) = \vec{z}_{NOS}]$. Figure 8.4 also shows that there are two additional sets of nodes to attack that are approximately as frequent. This alternative perspective implicitly suggests a heuristic approach to finding Robust Optimization Solutions.

Table 8.6 provides an alternative view of the same experiment. Here, for many randomly selected values of $\vec{a}(\omega)$ and $\vec{b}(\omega)$, we examine (i) the frequency that an attack vector solution has less impact on the network than the ROS, (ii) on average by how much, and (iii) how frequently the solution impairs the network to a level below 66.67. For example, as indicated in Table 8.6, empirically,

$$PR_\omega[\mathcal{R}(\vec{z}_{NOS}, \vec{a}(\omega), \vec{b}(\omega), \{\}) \le \mathcal{R}(\vec{z}_{ROS}, \vec{a}(\omega), \vec{b}(\omega), \{\})] = 35\%,$$

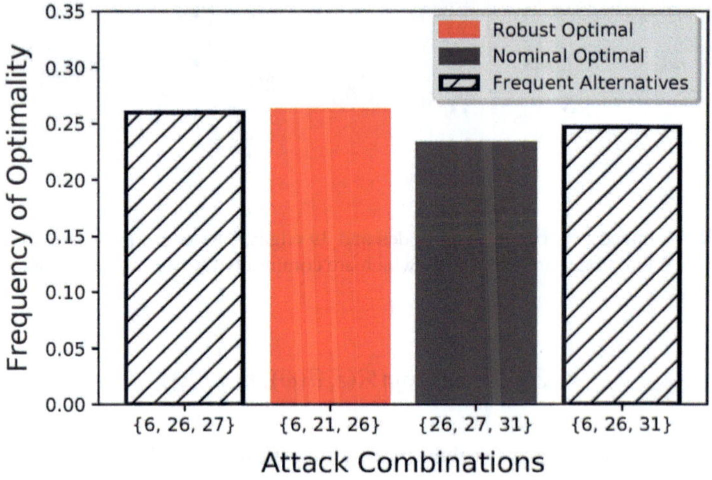

Fig. 8.4 The histogram of optimal solutions for 10,000 samples of $\Gamma_\alpha = \Gamma_\beta = 20$, $B_L = B_H = 3$ based on the network in Fig. 8.3. The Robust Optimal Solution is {6, 21, 26}. It is also the most likely optimal solution; however the margin of superiority is small

Table 8.6 A pairwise comparison of the solutions in Fig. 8.4 to the Robust Optimal Solution (ROS). The second column shows how often a solution has a higher (worse) solution than ROS. The third column shows the average increase in the system MOE across the samples (recall an increase indicates the attacker is less effective) compared with ROS. The last column shows a comparison of the solutions using $\frac{2}{3}$ of potential MOE as a cutoff value. This shows the probabilities of successfully reducing the MOE below the cutoff value

Solution	Freq. higher MOE than ROS	Avg. MOE increase over ROS	$\Pr[\mathcal{R}(\cdot) \leq 66.67]$
{6, 26, 27}	46%	0.9	49.3%
{6, 21, 26} (ROS)	NA	NA	74.3%
{26, 27, 31} (NOS)	35%	3.3	23.2%
{6, 26, 31}	42%	2.4	48.2%

$$mean_\omega\left(\mathcal{R}(\vec{z}_{NOS}, \vec{a}(\omega), \vec{b}(\omega), \{\}) - \mathcal{R}(\vec{z}_{ROS}, \vec{a}(\omega), \vec{b}(\omega), \{\})\right) = 3.3,$$

$$PR_\omega[\mathcal{R}(\vec{z}_{ROS}, \vec{a}(\omega), \vec{b}(\omega), \{\}) < 66.67] = 74.3\%,$$

and

$$PR_\omega[\mathcal{R}(\vec{z}_{NOS}, \vec{a}(\omega), \vec{b}(\omega), \{\}) < 66.67] = 23.2\%.$$

For this example, when comparing ROS with the other frequent solutions including NOS, if the attacker uses the ROS solution, the attacker will, on average, be slightly more effective (as the third column of Table 8.6 has only small positive numbers) and will be more likely to reduce the system to less than 66.67% (as the

Table 8.7 An example of three hypothetical trials of impact of ROS and NOS with the property that ROS, on average, is more impactful than NOS (50% vs. 60%), ROS is more likely than NOS to have a trial value of MOE less than 66.67% (2/3 vs. 1/3), yet ROS is more likely (2/3 vs. 1/3) to result in a higher (worse) MOE than NOS

	$PR[\mathcal{R}_{ROS}(\cdot) \leq \mathcal{R}_{NOS}(\cdot)]$	$PR[\mathcal{R}_{NOS}(\cdot) \leq \mathcal{R}_{ROS}(\cdot)]$
Trial 1	20%	80%
Trial 2	50%	30%
Trial 3	80%	70%

fourth column of Table 8.6 is higher in the fourth column for ROS) yet will be more likely than not to be *less* effective at reducing the MOE (as the second column of Table 8.6 only has entries less than 50%). This may appear to be an impossibly inconsistent finding, but it is not. Such conclusions are consistent with ROS being typically slightly less effective, whereas it is rarely radically less effective. This can be seen in the artificial example in Table 8.7 of three trials. Here, ROS, on average, is more effective than NOS (50% vs. 60%), ROS is more likely than NOS to have a trial value of MOE less than 66.67% (2/3 vs. 1/3), yet ROS is more likely (2/3 vs. 1/3) to result in a higher (worse) MOE than NOS. This occurs because ROS is slightly worse (higher) than NOS 2/3 of the times, but substantially better than NOS 1/3 of the times. This is the power of the robust analysis.

8.3.2 *Resilience Through Reconfiguration*

One possible way to increase the resilience of a mission network is to create the option of reconfiguring it using nodes that perform similar functions. Using a minor modification of the methods developed in this chapter, this section will show that dependency representations can easily implement a reconfiguration analysis that works with the Mission Network Pivot Algorithm and still provide robust solutions. Figure 8.5 shows a possible illustrative scenario where reconfiguration may increase the resilience of the network. Suppose node 7 performs an action that can support node 2 in a similar manner to the action that node 6 performs. Node 7 does not currently support node 2, but through reconfiguration, it could (as indicated by the dashed line). The right side of Fig. 8.5 shows what the reconfiguration would look like. Node 2 is now more resilient, in that if node 6 is attacked, node 2 will still receive support from node 7. It is possible to capture network reconfiguration by introducing an OR node. OR node values are governed by (8.33), which can replace (8.1) for all OR nodes in the network (Servi and Garvey 2017). Thus, network reconfiguration is handled simply by adding OR nodes for possible connections and using all the same methods of optimization and analysis. Note that if a non-FDNA dependency equation were used in place of (8.1), a different OR dependency equation should be constructed in a similar manner to model reconfiguration.

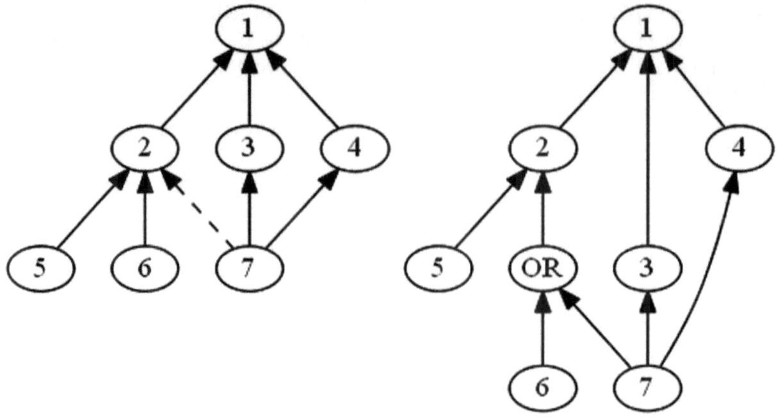

Fig. 8.5 Left: a network in which nodes 6 and 7 perform similar tasks, but node 7 does not engage with node 2. Right: a reconfiguration to allow node 7 to support node 2 in case node 6 fails or is attacked. Use of an OR node in the analysis captures such reconfigurations

$$P_i = \max_{n \in \mathcal{J}(m)} \left(\min \left[\alpha_{ij} P_j + 100 \left(1 - \alpha_{ij}\right), \left(P_j + \beta_{ij}\right) \right] \right). \tag{8.33}$$

8.4 Strategic Interactions Between the Defender and the Attacker

The framework in Sects. 8.1.3 and 8.1.4 identified the best decisions for the BLUE and RED, respectively, implicitly assuming that their opponents' decisions were fixed and known. In this section, we model the interactions between BLUE's decisions and RED's decisions on the same mission network. There are several interactions that we can model with the framework using the input variables ℓ_i and h_i defined in those sections. The variable ℓ_i captures the extent that an attack on the undefended node i is degraded, and the variable h_i captures the impact on node i of a defender mitigation investment.

Specifically, in this section we examine a two-player, two-stage game, often referred to as a Stackelberg Game, in which RED first attacks the network and then BLUE makes focused repairs and upgrades. To model the Stackelberg Game, we allow full knowledge of preferences and payoffs between RED and BLUE and allow them to act in sequence. Unlike the analysis in Sects. 8.1.3 and 8.1.4, here it is assumed that RED's actions will take into consideration BLUE's response. To examine how BLUE would mitigate an offensive strike from RED, we allow RED to act first and set $h_i > 1, \forall i \in N$. One may interpret $h_i > 1$ for any node i to mean that BLUE's investment upgrades node i rather than just bolstering or repairing it.

For this interaction we introduce the following notation: $\mathcal{G}_B(\vec{\ell}, \vec{h})$ is the defender (or BLUE) decisions, governed by (8.2)–(8.8), and $\mathcal{G}_R(\vec{\ell})$ is the attacker (or RED) decisions and governed by (8.9)–(8.16). The following analysis is generated using the network in Fig. 8.3.

For a small network, it is possible to visualize the preferences and payoffs in tabular format. Table 8.8 shows the values of $(\mathcal{G}_R(\vec{\ell}), \mathcal{G}_B(\vec{\ell}, \vec{h}))$ for each optimal attack by RED and subsequent repair by BLUE at each combination of quantity of nodes attacked and repaired, where $\psi^z = 5$ and $\psi^w = 10$ using the analysis from Sects. 8.1.3 and 8.1.4.

For each row of Table 8.8, BLUE's decision picks the column that maximizes $\mathcal{G}_B(\vec{\ell}, \vec{h})$, the second number, and the cells that correspond to those choices are highlighted in gray. The gray cells are BLUE's optimal responses to any of RED's starting choices. Knowing this, RED will choose from those eight optimal responses the attack, which minimizes $\mathcal{G}_R(\vec{\ell})$, i.e., RED will look at the first number of the highlighted cells and identify the smallest one. That combination is $(\mathcal{G}_R(\vec{\ell}), \mathcal{G}_B(\vec{\ell}, \vec{h})) = (82, 52)$, corresponding to RED attacking four nodes and BLUE repairing one node. By definition this is the Stackelberg Equilibrium solution. This cell is highlighted in yellow. It is interesting to note that this solution also has the property that (i) given RED's decision to attack four nodes, the best action of BLUE to maximize the second number in Table 8.8 is to repair one node, and (ii) given BLUE's decision to repair one node, the best action of RED to minimize the first number in Table 8.8 is to attack four nodes. Hence, by definition, for this example the Stackelberg Equilibrium solution is also the Nash Equilibrium solution.

By looking at the same interaction with different values for ψ^z and ψ^w, we can create an interesting phenomenon. If we let $\psi^z = \psi^w = 4$, we can show that the system is *antifragile*. Antifragility is the convex response of a system to stressors (Taleb 2012). Simply put, in some cases, for sufficiently small ψ^w, RED is worse off, if RED chooses to attack *any* nodes in the network, i.e., if RED accounts for the response of BLUE in its action by minimizing $\mathcal{G}_R(\vec{\ell})$, RED will choose not to attack any node. Antifragility is a powerful property for a network defender because it can act as a deterrent to attackers. Since RED understands the full impact of its attacks on the network and how BLUE will optimally respond, the antifragility property will deter RED from attacking the network at all. Therefore, if BLUE believes the system is antifragile, educating RED of this situation in order to deter any attacks is a form of deterrence. Table 8.9 shows the payoff table for this new interaction. The Stackelberg Equilibrium for RED and BLUE in this scenario (highlighted in yellow in Table 8.9) occurs when neither RED nor BLUE alters the network in anyway. This particular system is then antifragile because any possible harm done by RED to BLUE will result in an increase in $\mathcal{G}_R(\vec{\ell})$ after BLUE acts. Since this is not in RED's interest, it will choose not to attack any node.

This can be intuitively understood by considering the situation where investments by BLUE to harden a node result in the node being better than it would be if not attacked and then hardened, e.g., $h_i > 1$. Here, BLUE might choose not to upgrade its network, but if forced to due to an attack by RED, it will be better off than before the attack. This antifragility property serves as a deterrent for RED to attack.

Table 8.8 The payoff table for a Stackelberg Game between RED and BLUE where $\psi^z = 5$ and $\psi^w = 10$ based on the mission network represented by Fig. 8.3, where each row corresponds to the number of nodes attacked, each column is the number of nodes repaired, and each entry corresponds to the optimal objective value for the attacker, RED, and the defender, BLUE, $(\mathcal{G}_R(\vec{\ell}), \mathcal{G}_B(\vec{\ell}, \vec{h})$. The cells highlighted in gray correspond to all possible optimal numbers of nodes for BLUE to repair based on the number of nodes attacked by RED. The yellow cell corresponds to the Stackelberg Equilibrium, i.e., the lowest objective for BLUE among the highlighted cells

		Number of nodes repaired							
		0	1	2	3	4	5	6	7
Number of nodes attacked	0	82.6; 82.6	83.6; 73.6	84.4; 64.4	85.0; 55.0	85.5; 45.5	86.0; 36.0	86.3; 26.3	86.5; 16.5
	1	77.4; 72.4	88.6; 73.6	89.4; 64.4	90.0; 55.0	90.5; 45.5	91.0; 36.0	91.3; 26.3	91.5; 16.5
	2	74.3; 64.3	85.5; 65.5	94.3; 64.3	95.0; 55.0	95.5; 45.5	96.0; 36.0	96.3; 26.3	96.5; 16.5
	3	71.7; 56.7	82.9; 57.9	91.7; 56.7	100.0; 55.0	100.5; 45.5	100.9; 36.0	101.2; 26.3	101.4; 16.5
	4	70.9; 50.9	82.0; 52.0	90.9; 50.9	99.2; 49.2	105.0; 45.0	105.5; 35.5	106.0; 26.0	106.4; 16.5
	5	71.0; 46.0	82.3; 47.3	91.1; 46.1	99.4; 44.4	105.2; 40.2	110.5; 35.5	111.0; 26.0	111.4; 16.5
	6	71.5; 41.5	82.7; 42.7	91.5; 41.5	99.8; 39.8	105.6; 35.7	110.7; 30.7	115.9; 26.0	116.4; 16.5
	7	73.6; 38.6	84.8; 39.8	93.6; 38.6	101.9; 36.9	107.7; 32.8	113.0; 28.1	118.1; 23.1	121.0; 16.0

Table 8.9 The payoff table for a Stackelberg Game between RED and BLUE, where $\psi^z = \psi^w = 4$ based on the mission network represented by Fig. 8.3, with each table entry equaling $(\mathcal{G}_R(\vec{\ell}), \mathcal{G}_B(\vec{\ell}, \vec{h}))$ based on the number of nodes attacked and repaired. The cells highlighted in gray correspond to all possible optimal responses for BLUE. The yellow cell corresponds to the Stackelberg Equilibrium. Since RED is worse off if RED attacks any nodes, this network might be considered antifragile

		Number of nodes repaired							
		0	1	2	3	4	5	6	7
Number of nodes attacked	0	82.6; 82.6	83.6; 79.6	84.4; 76.4	85.0; 73.0	85.5; 69.5	86.0; 66.0	86.3; 62.3	86.5; 58.5
	1	76.4; 72.4	87.6; 79.6	88.4; 76.4	89.0; 73.0	89.5; 69.5	90.0; 66.0	90.3; 62.3	90.5; 58.5
	2	72.3; 64.3	83.5; 71.5	92.3; 76.3	93.0; 73.0	93.5; 69.5	94.0; 66.0	94.3; 62.3	94.5; 58.5
	3	68.7; 56.7	79.9; 63.9	88.7; 68.7	97.0; 73.0	97.5; 69.5	98.0; 66.0	98.3; 62.3	98.5; 58.5
	4	66.9; 50.9	78.1; 58.1	86.9; 62.9	95.2; 67.2	101.0; 69.0	101.5; 65.5	102.0; 62.0	102.4; 58.5
	5	66.1; 46.1	77.3; 53.3	86.1; 58.1	94.4; 62.4	100.2; 64.2	105.5; 65.5	106.0; 62.0	106.4; 58.5
	6	65.5; 41.5	76.7; 48.7	85.5; 53.5	93.8; 57.8	99.6; 59.6	104.7; 60.7	109.9; 62.0	110.4; 58.5
	7	66.6; 38.6	77.8; 45.8	86.6; 50.6	94.9; 54.9	100.7; 56.8	106.0; 58.0	111.1; 59.1	114.0; 58.0

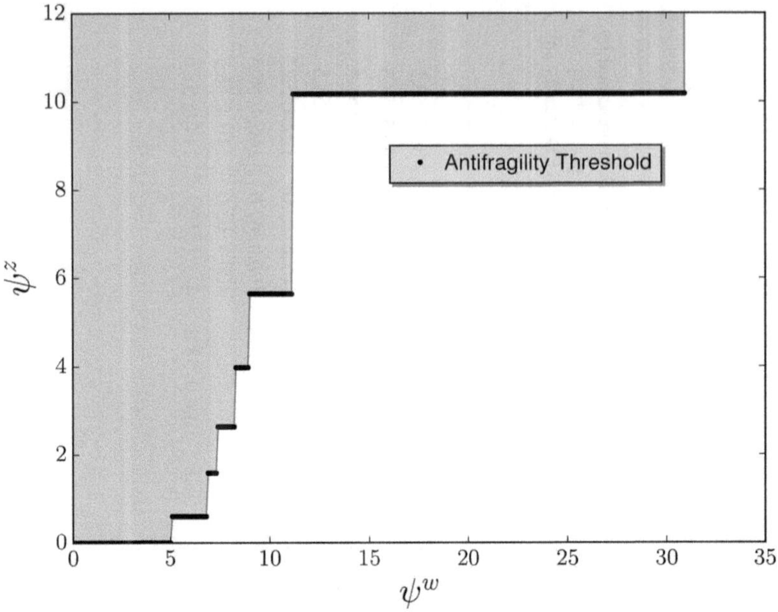

Fig. 8.6 The regions of antifragility based on ψ^z and ψ^w. The gray shaded area yields interaction models where no attacks are made on the network. Analysis is based on the network in Fig. 8.3

To be clear, it is not the network that is inherently antifragile, but rather that the interaction between RED and BLUE with their respective inclination to alter the network (as reflected by ψ) which makes the network antifragile. We can examine the ψ^z thresholds for which the system is antifragile given a known ψ^w. Figure 8.6 shows the upper limit on ψ^z that will result in RED attacking the network for varying levels of ψ^w. Combinations in the gray shaded region yield an antifragile system, and RED will choose not to attack the network.

8.5 Conclusion

This chapter creates a framework for robust mission network analysis using FDNA. The general formulations showed how a user could choose the best attack or defense strategy while balancing the investments costs with the impact on the Measure of Effectiveness. The general formulations also allowed tuning of two parameters, h_i and ℓ_i, to change the scenario being modeled from one of protection to one of repair and mitigation. The new algorithm presented searches for and bounded the Robust Optimization Solution using the valuation method chosen to model the mission network. The algorithm can be used with any mission network model that can be formulated as a maximin (see (8.17)–(8.27)) and was illustrated with an example.

The initial analytical framework encountered a nonlinearity which would slow down the optimization. However, this chapter showed how to circumvent this nonlinearity. Using random scenarios, this chapter compared the robust optimal solution with other solutions. This chapter also showed how the process could easily handle network reconfiguration with a small addition of the OR node (see (8.33)). Lastly, game theory was introduced in Sect. 8.4 and was used to show ways the models could be combined to capture the strategic interactions between an attacker and a defender (i.e., finding RED's optimal attack enabled BLUE to react and improve the network) and the possibility of antifragility properties, which might serve as a deterrent to RED.

Appendix: Network Characterization

Figure 8.3 is a randomly generated network with the baseline operabilities in Table 8.10 and the dependency parameters outlined in Table 8.11. We chose this network because it is small enough for clarity of demonstration and it is complex enough to demonstrate the methods.

Table 8.10 The baseline operating value for each leaf node in the network in Fig. 8.3. In the absence of attacks or investments, these are the MOE values for each leaf node

ID	5	6	8	11	12	14	16	17	19	20	21	24	25	26
Operability	67	92	46	50	99	42	86	89	57	73	46	81	64	64

ID	27	28	29	30	31	32	33	34	35	36	37	38	39
Operability	75	96	61	89	55	90	89	42	97	74	48	70	96

Table 8.11 The α and β parameters for each edge that connects the parent node to the child node listed for Fig. 8.3

Parent	0	0	0	0	0	0	0	1	1	1	2	2	2
Child	1	2	3	4	9	27	31	6	19	20	17	18	35
α	0.64	0.89	0.43	0.46	0.73	0.95	0.96	0.8	0.89	0.96	0.33	0.12	0.36
β	94	63	82	72	81	78	81	11	18	42	41	11	61

Parent	2	3	3	3	3	4	4	4	4	4	4	7	7
Child	36	7	11	12	14	5	8	10	28	34	38	16	37
α	0.46	0.12	0.34	0.25	0.97	0.92	0.07	0.85	0.97	0.26	0.48	0.35	0.12
β	75	28	34	57	63	74	38	57	38	99	7	86	11

Parent	9	9	9	10	10	13	15	18	18	18	22	23	23
Child	13	21	23	15	22	33	32	26	29	30	39	24	25
α	0.2	0.06	0.98	0.7	0.89	0.84	0.63	0.55	0.58	0.74	0.14	0.14	0.55
β	95	39	68	80	47	23	28	15	66	63	6	16	59

References

Bertsimas D, Sim M (2004) The price of robustness. Oper Res 52(1):35–53
Child J (2000) Julia's kitchen wisdom: essential techniques and recipes from a lifetime of cooking. Alfred A. Knopf, New York
McCormick GP (1976) Computability of global solutions to factorable nonconvex programs: Part I—convex underestimating problems. Math Program 10(1):147–175
Garvey PR, Pinto CA (2009) Introduction to functional dependency network analysis. In: Proceedings of the second international symposium on engineering systems. Massachusetts Institute of Technology, Cambridge, MA
Servi LD, Garvey PR (2017) Deriving global criticality conditions from local dependencies using functional dependency network analysis (FDNA). Syst Eng 20(4):297–306
Guariniello C, DeLaurentis D (2017) Supporting design via the System Operational Dependency Analysis methodology. Res Eng Des 28(1):53–69
Birge JR, Louveaux F (2011) Introduction to stochastic programming, 2nd edn. Springer, Berlin
Prékopa A (2013) Stochastic programming. mathematics and its applications, vol. 324. Springer Science & Business Media, Berlin
Ben-Tal A (1985) The entropic penalty approach to stochastic programming. Math Oper Res 10(2):263–279
Taleb NN (2012) Antifragile: things that gain from disorder. Random House, New York
Watters J, Morrissey S, Bodeau D, Powers S (2009) The Risk-to-Mission Assessment Process (RiskMAP): a sensitivity analysis and an extension to treat confidentiality issues. In: The MITRE Corporation/Institute for Information Infrastructure Protection (I3P), Pub Rel 09-2994

Chapter 9
Incorporating Time into Mission Dependency Network Analysis

'Time,' he said,' is what keeps everything from happening at once.'

Ray Cummings (1919)

9.1 Introduction

The aim of this chapter is to introduce a new model for mission network analysis, where the operability of any node can vary with time. There are several existing frameworks for analyzing dependency representations such as FDNA and SODA, which were introduced and used throughout this book. These frameworks have been studied and can be used to identify critical nodes in the mission network, but such analysis is typically performed in a static fashion. In many applications, immediate system failure is not the only concern to a mission, and analyses should also account for the time it takes to detect the failure and recover the system operability.

The model presented in this chapter will be formulated in terms of a hybrid mission network. A hybrid mission network consists of nodes that are designated as either *physical* or *functional*. This class of networks can naturally be used to model missions that are comprised of a sequence of interrelated activities (i.e., functional nodes), where the ability to complete these activities relies on the operability of physical systems or infrastructure (i.e., physical nodes).

After introducing notations, Sect. 9.2 will introduce, in steps, a new model for mapping mission network dependencies that can account for time-varying effects. This captures physical node detection and recovery times, as well as transient effects for functional node operability. The model will first be formulated as an optimization problem from an attacker's point of view, to determine which physical nodes are best to target. Next, we consider a set of defensive actions to reduce the detection time for physical node failure and then pose a bi-level optimization problem in which the defender can best fortify their network while anticipating the attacker's adaptation.

L. Servi, *Mission Dependency Network Analysis*, Risk, Systems and Decisions,
https://doi.org/10.1007/978-3-032-11260-6_9

Specifically, Sect. 9.2 frames the basic model of a hybrid mission network that can have physical nodes attacked, and all ensuing effects are first considered to be instantaneous, before being formulated with a time-delayed response to each functional node's operability to the attacks. Next, in Sect. 9.3, the time-varying response to the physical nodes is captured first in an exact form and then in an approximate form that handles the potential complication of one attack occurring before the network has completely recovered from a previous attack. Section 9.4 introduces the defender's decision to lessen the time to detect and start the recovery of physical node failures using a heuristic algorithm to find the Stackelberg Equilibrium of the combined attacker-defender formulation. Section 9.5 ends with an illustrative time-varying example.

This chapter has four appendices: The first formally states the precise bi-level optimization problem. The second provides more details regarding the response function for the case of overlapping physical node recoveries. The third illustrates the impact on the best defense strategy, assuming the functional node operability changes in time when its dependent physical node's operability changes. Finally, the fourth appendix computes an upper bound for the error of the recovery function approximation introduced in Sect. 9.3.

9.1.1 Notation

We use the following notation in our initial formulation for the hybrid mission network analysis model.

- START node = the single functional node that is assumed to be fully operable and to have no physical nodes supporting it. It is the sole source of all paths reaching all other functional nodes in the network
- n_i = the number of physical nodes supporting functional node i
- $\vec{P}_i(t) = (P_{it}^1, P_{it}^2, \ldots, P_{it}^{n_i})$ the operabilities of the n_i physical nodes, $(i, 1), (i, 2), \ldots, (i, n_i)$, supporting functional node i
- $\mathcal{P}_i$ = the set of paths from START to functional node i
- $\mathcal{F}_i$ = the set of functional nodes that directly support/precede functional node i. Hence, $\mathcal{F}_i$ fully defines the network topology of the functional nodes. For example, if each functional node is supported by exactly one other functional node, then $\mathcal{F}_i = \{i - 1\}$ for all i.
- $P_{it}^0 = F_i\left(\vec{P}_i(t)\right)$, an upper limit of the operability of functional node i at time t, based on the operability of physical nodes, $(i, 1), (i, 2), \ldots, (i, n_i)$, that support functional node i. This might be through an FDNA-like (Garvey et al. 2014) and SODA-like (Guariniello and DeLaurentis 2017) relationships such as

$$
\begin{aligned}
P_{it}^{0} &= F_i\left(\vec{P}_i(t)\right) \\
&= \begin{cases} \min\left[\min_{j\in\{1,\ldots,n_i\}}[\gamma_{ji}P_{it}^{j}+\beta_{ji}],\, \sum_{j=1}^{n_i}\alpha_{ji}P_{it}^{j}+\delta_i\right], & i \neq START, \\ 100, & i = START, \end{cases}
\end{aligned}
\tag{9.1}
$$

where either min can be replaced with a max.

- $\phi_{i,t}$ = the operability of functional node i at time t. This is given by

$$
\phi_{i,t} = \begin{cases} \min\left[P_{it}^{0},\, \max_{\ell\in\mathcal{F}_i}[\phi_{\ell t}]\right] = \max_{p\in\mathcal{P}_i}\left[\min_{k\in p}[P_{kt}^{0}],\right], & \text{if } i \neq \text{START}, \\ 100, & \text{if } i = \text{START}. \end{cases}
\tag{9.2}
$$

If i is not the START node, the first definition can be interpreted as assuming that the operability of functional node i at time t is constrained to be no more than the operability of its preceding functional nodes. A recursive argument can be used to see that this is equivalent to assuming that the operability of functional node i at time t can be found by identifying the path from START to node i, whose minimum value of P_{it}^{0} is the largest. If i is the START node, the operability of the node is, by convention, 100.

This recurrence relation says that the operability of the best path to functional node i is equal to the operability of the best path to any node that precedes functional node i or the operability of functional node i, whichever is lower.

If $i \neq START$, substituting the first equation in (9.1) into (9.2) yields

$$
\begin{aligned}
\phi_{it} &= \min\left[\min\left[\min_{j\in\{1,\ldots,n_i\}}[\gamma_{ji}P_{it}^{j}+\beta_{ji}],\, \sum_{j=1}^{n_i}\alpha_{ji}P_{it}^{j}+\delta_i\right],\, \max_{\ell\in\mathcal{F}_i}[\phi_{\ell t}]\right] \\
&= \min\left[\min_{j\in\{1,\ldots,n_i\}}[\gamma_{ji}P_{it}^{j}+\beta_{ji}],\, \sum_{j=1}^{n_i}\alpha_{ji}P_{it}^{j}+\delta_i,\, \max_{\ell\in\mathcal{F}_i}[\phi_{\ell t}]\right].
\end{aligned}
\tag{9.3}
$$

9.2 A First Optimization Model

In this section, we construct an optimization model to capture the behavior of a network after certain physical nodes have been disabled. This optimization model can be used to determine the subset of physical nodes to disable for the highest impact on overall network operability, subject to some budgetary constraint.

We begin with defining the relevant notations for the optimization model:

User-Defined Input Parameters:

- $\alpha_{ji}, \beta_{ji}, \gamma_{ji}, \delta_i$: network dependency parameters relating the functional node operability to physical node operability introduced in (9.1)
- B: the adversary's budget. This controls how many physical nodes the optimization will disable.
- T_{ij}: the maximum operability of physical node (i, j) if it is not disabled
- b_{ij}: the impact of disabling physical node (i, j). If disabled, physical node (i, j) will have an operability of $T_{ij}b_{ij}$.
- c_{ij}: the cost to disable physical node (i, j)
- ρ_i: the speed that the system responds to change. The parameter ρ_i is first implicitly defined in (9.12).
- w_t: the weight of the objective at time t to capture the relative importance of the timing of when the mission is impacted
- $\mathcal{F}_i$: the set of functional nodes that directly support/precede functional node i (as defined in Sect. 9.1.1)

Decision Variables

- z_{ijt}: the binary decision which denotes whether or not to decise to disable physical node (i, j), supporting functional node i, at time t. Setting z_{ijt} to 1 corresponds to deciding to disable the node. Otherwise, it is set to 0.
- M: a large number needed for the optimization

Intermediate Variables Used

- P_{it}^0 is the operability of functional node i at time t.
- P_{it}^j is the operability of physical node (i, j), supporting functional node i, at time t.
- $\phi_{i,t}$ is the value of the best path at time t from the functional START node to functional node i.
- $\phi_{*,t}$ is the operability of the final functional node in the network.
- x_{ijt}^1 is an indicator variable to help compute $\phi_{i,t}$. x_{ijt}^1 equals 1 if $\phi_{i,t} = \gamma_{ji} P_{it}^j + \beta_{ji}$ and 0 otherwise.
- x_{it}^2 is an indicator variable to help compute $\phi_{i,t}$. x_{it}^2 equals 1 if $\phi_{i,t} = \sum_j \alpha_{ji} P_{it}^j + \delta_i$ and 0 otherwise.
- x_{it}^3 is an indicator variable to help compute $\phi_{i,t}$. x_{it}^3 if $\phi_{i,t} = \max_{\ell \in \mathcal{F}_i}[\phi_{\ell t}]$ and 0 otherwise.

9.2.1 Instant Dynamics

First, we consider the case where each time period is in a distinct "sandbox," and the nodal failures from one time period do not affect the operability values in any subsequent time periods. This situation can be described with the following model:

Minimize

$$\sum_t w_t \phi_{*,t} \tag{9.4}$$

subject to

$$\phi_{i,t} \geq \gamma_{ji} P_{it}^j + \beta_{ji} - (1 - x_{ijt}^1)M, \qquad \forall i, j, t, \tag{9.5}$$

$$\phi_{i,t} \geq \sum_{j=1}^{n_i} \alpha_{ji} P_{it}^j + \delta_i - (1 - x_{it}^2)M, \qquad \forall i, t, \tag{9.6}$$

$$\phi_{i,t} \geq \phi_{\ell t} - (1 - x_{it}^3)M, \qquad \forall i, t \ \forall \ell \in \mathcal{F}_i, \tag{9.7}$$

$$1 = \sum_{j=1}^{n_i} x_{ijt}^1 + x_{it}^2 + x_{it}^3, \qquad \forall i, t, \tag{9.8}$$

$$P_{it}^j \geq T_{ij}[1 - z_{ijt}(1 - b_{ij})], \qquad \forall (i, j), t, \tag{9.9}$$

$$B \geq \sum_{i,j,t} c_{ij} z_{ijt}, \tag{9.10}$$

$$z, x \text{ binary}. \tag{9.11}$$

The model objective above, given in (9.4), sums the operability of the final functional node of the network at each time step, weighted according to w_t. These weights can be used to discount future timesteps or to place higher emphasis on a particular time range of interest.

Constraints (9.5)–(9.8) are used to compute the value of $\phi_{i,t}$. From (9.8), exactly one of these constraints will be active, corresponding to which of the arguments in (9.3) attains the minimum. The exact argument will be indicated by variable x, i.e., either x^1, x^2, or x^3 will be 1. This takes advantage of the fact that the optimization is aiming to minimize a linear combination of the $\phi_{*,t}$'s.

Constraint (9.9) is used to determine the operability of physical node (i, j) at time t, based on the binary attack vector z_{ijt}. If physical node (i, j) is attacked at time t, i.e., $z_{ijt} = 1$, then it has operability $P_{it}^j = T_{ij} b_{ij}$. Otherwise, it has operability $P_{it}^j = T_{ij}$. Here, T_{ij} and b_{ij} are physical node parameters in [0, 100] and [0, 1], respectively. Constraint (9.10) is the budget restricting the total number of attacks that can be made.

9.2.2 Delayed Response

If the operability is a continuous time variable, it is natural to introduce it evolving in time. Assume that functional node operability does not change instantaneously, as in (9.2), but instead follows the dynamics:

$$\frac{\phi_{i,t+\Delta} - \phi_{i,t}}{\Delta} = \rho_i \left(\min \left[P^0_{i,t+\Delta}, \max_{\ell \in \mathcal{F}_i} [\phi_{\ell,t})] \right] - \phi_{i,t} \right), \tag{9.12}$$

where ρ_i is another parameter for node i to control how quickly changes in operability propagate. Specifically, (9.12) is a generalization of (9.2) to a dynamic case. Here, rather than assuming that the left-hand side of (9.2) instantaneously equals the right-hand side, the right-hand side dynamically evolves toward the left-hand side value at a rate of ρ_i. Equation (9.12) gives rise to the following update rule:

$$\phi_{i,t+\Delta} = (1 - \rho_i \Delta)\phi_{i,t} + \rho_i \Delta \min \left[P^0_{i,t+\Delta}, \max_{\ell \in \mathcal{F}_i} [\phi_{\ell,t}] \right]. \tag{9.13}$$

In this case, to adjust the previous optimization model, we bring the update rule (9.13) directly into the constraints (9.5)–(9.7). Specifically, (9.13) and (9.1) imply

$$\phi_{i,t+\Delta} = (1 - \rho_i \Delta)\phi_{i,t} + \rho_i \Delta \min \left[\min_{j \in \{1,\ldots,n_i\}} [\gamma_{ji} P^j_{i,t+\Delta} + \beta_{ji}], \sum_{j=1}^{n_i} \alpha_{ji} P^j_{i,t+\Delta} \right.$$
$$\left. + \delta_i, \max_{\ell \in \mathcal{F}_i} [\phi_{\ell,t+\Delta}] \right], \tag{9.14}$$

which captures the dynamic evolution of the operability of functional node i at time $t + \Delta$, denoted as $\phi_{i,t+\Delta}$, in terms of its previous value, $\phi_{i,t}$, the operability of the dependent physical nodes at time $t + \Delta$, and the functional nodes which directly support functional node i at time $t + \Delta$. Since all terms are nonnegative, this can be rewritten as

$$\phi_{i,t+\Delta} = \min \left[\min_{j=1,\ldots,n_i} \left[(1 - \rho_i \Delta)\phi_{i,t} + \rho_i \Delta \left(\gamma_{ji} P^j_{i,t+\Delta} + \beta_{ji} \right) \right], (1 - \rho_i \Delta)\phi_{i,t} \right.$$
$$+ \rho_i \Delta \left(\sum_{j=1}^{n_i} \alpha_{ji} P^j_{i,t+\Delta} + \delta_i \right),$$
$$\left. \max_{\ell \in \mathcal{F}_i} \left[(1 - \rho_i \Delta)\phi_{i,t} + \rho_i \Delta \phi_{\ell,t+\Delta} \right] \right]. \tag{9.15}$$

This gives rise to the following revised set of constraints. Below, we substitute Δ for 1 in (9.15) to find that

$$\phi_{i,t} \geq (1-\rho_i)\phi_{i,t-1} + \rho_i \left(\gamma_{ji} P^j_{it} + \beta_{ji}\right) - (1 - x^1_{ijt})M, \qquad \forall i, j, t > 0, \tag{9.16}$$

$$\phi_{i,t} \geq (1-\rho_i)\phi_{i,t-1} + \rho_i \left(\sum_{j=1}^{n_i} \alpha_{ji} P^j_{it} + \delta_i\right) - (1 - x^2_{it})M, \qquad \forall i, t > 0, \tag{9.17}$$

$$\phi_{i,t} \geq (1-\rho_i)\phi_{i,t-1} + \rho_i \phi_{\ell,t} - (1 - x^3_{it})M, \qquad \forall i, t > 0, \ell \in \mathcal{F}_i. \tag{9.18}$$

These constraints can serve to replace those given by (9.5)–(9.7). For $t = 0$, the old constraints would apply. Here, M is a positive constant large enough so that when one of the x variables is on, the right-hand side of the corresponding equation is at most zero. This allows the x variables to act as a switch, effectively determining which argument is the minimum on the right-hand side of (9.15). This is done by adding the additional sets of constraints:

$$\sum_{j=1}^{n_i} x^1_{ijt} + x^2_{it} + x^3_{it} = 1, \qquad \forall i, t > 0. \tag{9.19}$$

9.3 Incorporating Physical Node Recovery

The formulation in Sect. 9.2.1 assumes that physical nodes are "down" in the period that they are attacked, but they recover back to full operability in the next period. In practice, an attack on a physical node may have a transient effect, where node operability may be impacted for several time periods after the attack takes place. We aim to reformulate the model to enforce this kind of behavior. First, we discuss these modifications for a general response function for physical node operability after an attack. We then propose a simple parameterized response function that can be used for analysis.

9.3.1 Physical Node Response

For $t \geq 0$, let $R_{ij}(t)$ represent the operability of a physical node (i, j), t units of time after it has been attacked. We assume that the response function for each physical node is known a priori and can therefore be treated as data in the model. In the case of the network dynamics presented in Sect. 9.2.1,

$$R_{ij}(t) = \begin{cases} T_{ij} b_{ij}, & t = 0 \\ T_{ij}, & \text{otherwise.} \end{cases}$$

More realistically, and depending on the granularity of the timescale, after the operability reaches a minimum, it may remain there temporarily and slowly regain operability over several time steps.

By definition of $R_{ij}(t)$, if the physical node (i, j) was attacked at time τ, for $t \geq \tau$, its operability will be given by $P_{it}^{j} = R_{ij}(t - \tau)$. This relationship is valid for a single attack, but when multiple attacks occur on the same node, there is some ambiguity in how the operability should react. We make the following two assumptions about the attack response:

- $R_{ij}(t)$ has the property that the initial drop in operability happens instantaneously at $t = 0$, and the node recovers over subsequent time periods.
- Multiple attacks on the same node do not have compound effects on operability. Instead, we assume that the current node operability is only affected by the most recent attack.

This would imply that the operability of physical node (i, j) at time t, where $t > \tau$, can be written as

$$P_{it}^{j} = T_{ij} + \min_{\tau \leq t} \left[(R_{ij}(t - \tau) - T_{ij}) z_{ij\tau} \right]. \tag{9.20}$$

This condition can be enforced by introducing a series of binary variables, $W_{ijt\tau}$, to indicate which element in the right-hand side of (9.20) attains the minimum. That is,

$$P_{it}^{j} \geq T_{ij}(1 - z_{ij\tau}) + R_{ij}(t - \tau) z_{ij\tau} - (1 - W_{ijt\tau}) M, \quad \forall (i, j), t, \tau \leq t, \tag{9.21}$$

$$\sum_{\tau \leq t} W_{ijt\tau} = 1, \qquad \forall (i, j), t. \tag{9.22}$$

Constraints (9.21) and (9.22) can replace constraint (9.9) in the initial formulation. Here, M is a constant value greater than or equal to T_{ij}, which ensures that when $W_{ijt\tau} = 0$, the right-hand side of (9.21) is equal to zero. In the case where no attacks occur prior to t, then $z_{ij\tau} = 0$ for all τ, and the constraints dictate that $P_{it}^{j} \geq T_{ij}$. Instead, when $z_{ij\tau} = 1$ for multiple $\tau \leq t$, it will be most beneficial to the objective if $W_{ijt\tau^*} = 1$, where τ^* is the time of the most recent attack. This is because $R_{ij}(t - \tau^*) \leq R_{ij}(t - \tau)$ for any other attack time τ; as such this value minimizes the right-hand side of (9.21).

Note that there is likely a maximum timespan, after which $R_{ij}(t)$ returns to maximum operability T_{ij}. That is, there may exist t_{ij}^{max} such that $R_{ij}(t) = T_{ij}$ for all $t > t_{ij}^{max}$, which would not be materially relevent to compute (9.20) or (9.21)

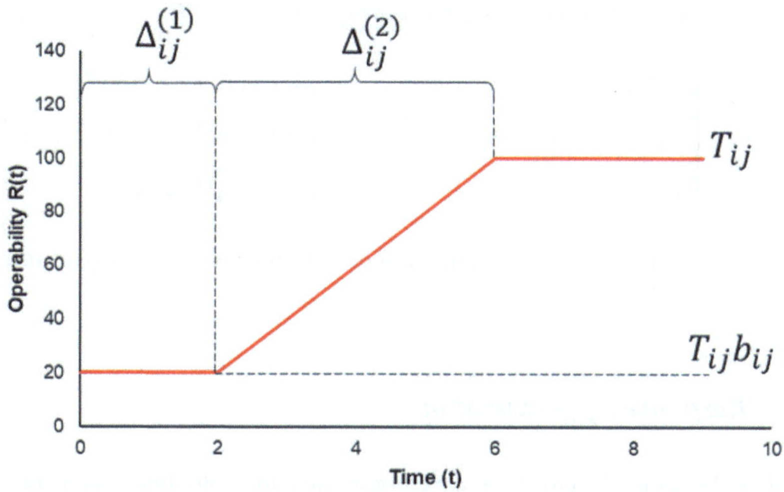

Fig. 9.1 An example response function where $\Delta_{ij}^{(1)} = 2$ and $\Delta_{ij}^{(2)} = 4$

and (9.22). This observation reduces the number of variables and constraints to only consider time periods τ between $\max[0, t - t_{ij}^{max}]$ and t. However, this method still increases the number of variables and constraints by

$$|\mathcal{T}| \sum_{i} \sum_{j=1}^{n_i} t_{ij}^{\max},$$

where $|\mathcal{T}|$ is the total number of time periods under consideration. In Sect. 9.3.3, we will discuss an approximation to these relationships that requires fewer variables and constraints.

9.3.2 *A Simple Class of Response Functions*

In this section, we propose a simple class of response functions that is more realistic than the instantaneous dynamics of Sect. 9.2.1 but is still simple enough to analyze. This family will rely on two parameters, $\Delta_{ij}^{(1)}$ and $\Delta_{ij}^{(2)}$.

We assume that each physical node goes through two distinct stages after being attacked. First, the node remains at its minimum operability for $\Delta_{ij}^{(1)}$ time periods. This could represent the time it takes to notice that a system is down. Then, it requires $\Delta_{ij}^{(2)}$ time periods to reach full operability again. We assume that in this phase, the physical node operability improves linearly. Figure 9.1 illustrates this behavior.

This behavior is captured with the following definition of R_{ij}:

$$R_{ij}(t) = \begin{cases} T_{ij} b_{ij}, & \text{for } t = 0, 1, \ldots, \Delta_{ij}^{(1)}, \\ T_{ij} b_{ij} + T_{ij}(1 - b_{ij}) \left\lceil \frac{t - \Delta_{ij}^{(1)}}{\Delta_{ij}^{(2)}} \right\rceil, & \text{for } t = \Delta_{ij}^{(1)}, \ldots, \Delta_{ij}^{(1)} + \Delta_{ij}^{(2)}, \\ T_{ij}, & \text{for } t \geq \Delta_{ij}^{(1)} + \Delta_{ij}^{(2)}. \end{cases} \tag{9.23}$$

For the remainder of this chapter, we focus on this specific family of response curves.

9.3.3 *Response Approximation*

Encoding the exact behavior of a response function into linear constraints as described in Sect. 9.3.1 requires the addition of several new constraints and binary variables to a Mixed Integer Linear Program. Depending on the adversarial budget, many physical nodes might only be attacked once or not at all. This implies that many of the constraints in (9.21) and (9.22) will be duplicative and unnecessary. To prevent this, we offer a slightly modified set of constraints that require fewer variables at the cost of some accuracy in describing the response function.

In lieu of constraints (9.21) and (9.22), consider the constraints:

$$P_{it}^{j} \geq T_{ij} - \sum_{\tau \leq t} \left(T_{ij} - R_{ij}(t - \tau) \right) z_{ij\tau} \qquad \forall\, (i, j), t. \tag{9.24}$$

One may interpret (9.24) as stating that the operability will be the maximum operability, T_{ij}, minus the summation of the unfinished restorative response for *each*, rather than just the last, disabling attacks.

With these constraints, if $z_{ijt} = 0$ for all t, they imply that $P_{it}^{j} \geq T_{ij}$. Likewise, if $z_{ij\tau} = 1$ for a single value of τ, the right-hand side of (9.24) will simplify to

$$P_{it}^{j} \geq T_{ij} - (T_{ij} - R_{ij}(t - \tau)) = R_{ij}(t - \tau),$$

which is the desired behavior. Also note that the summand of (9.24) vanishes for any $\tau < t - t_{ij}^{max}$ (in the case of our simple response functions, $t_{ij}^{max} = \Delta_{ij}^{(1)} + \Delta_{ij}^{(2)}$). Therefore, these constraints will also accurately describe multiple attacks on the same node, provided that they occur sufficiently far apart in time.

The only remaining case to consider is when multiple attacks occur on the same node that are close together in time. It is in this case where these simplified constraints represent an approximation to the ideal response function. We illustrate this approximation with the response function illustrated in Sect. 9.3.2, where $\Delta_{ij}^{(1)} = 2$ and $\Delta_{ij}^{(2)} = 4$. For simplicity, we also assume that $T_{ij} = 100$ and $b_{ij} = 0$.

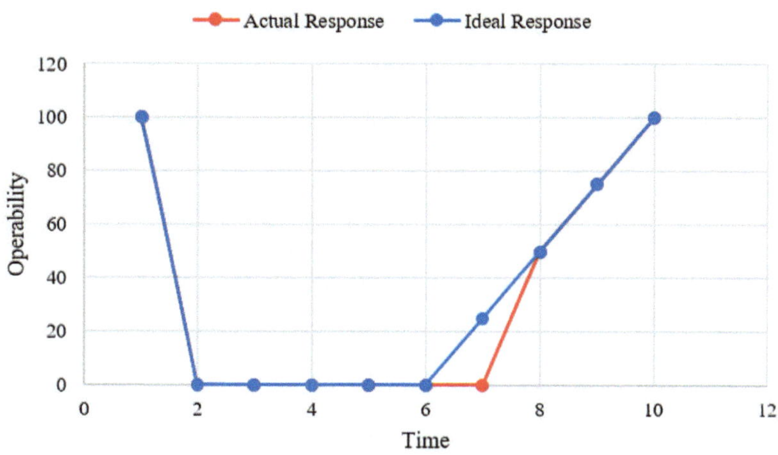

Fig. 9.2 The physical node response curve, if node (ij) was attacked at times 2 and 4. Blue represents the ideal reaction, and red represents the actual behavior enforced by the constraints

Let us consider what happens at times $t = 7$ if the node was attacked at times $t = 2$ and $t = 4$. Ideally, only the latter attack would register, and we would have $P^j_{i7} = R_{ij}(7-4) = 25$. Instead, following Eq. (9.24), we have

$$\begin{aligned} P^j_{i7} &\geq T_{ij} - [T_{ij} - R_{ij}(7-4)] - [T_{ij} - R_{ij}(7-2)] \\ &= 100 - (75) - (25) = 0. \end{aligned}$$

This deviation can be seen in Fig. 9.2, where both the ideal response (according to (9.21) and (9.22)) and the actual response (according to (9.24)) are plotted for times 0 through 10.

This issue with the constraints is not unique to the set of attacks chosen. If, instead, node (i, j) was attacked at times 2 and 3, the response curve will follow that seen in Fig. 9.3. Again, we see that in time periods 6 and 7, the actual response enforced by these constraints differs from the ideal response.

Although this approach is not perfect, it still provides a reasonable approximation to the response curve in virtually all scenarios. Therefore, we view this minor loss of accuracy as an acceptable trade-off for the relative gains in computational tractability. Furthermore, the response functions specified in Sect. 9.3.2 are still a somewhat simplistic way modeling node failure and recovery. Since they serve as a gross approximation to reality, it does not make sense at this point to make the model overly complex to fit their behavior exactly. Going forward, we will opt to use constraint (9.24) in lieu of (9.21) and (9.22).

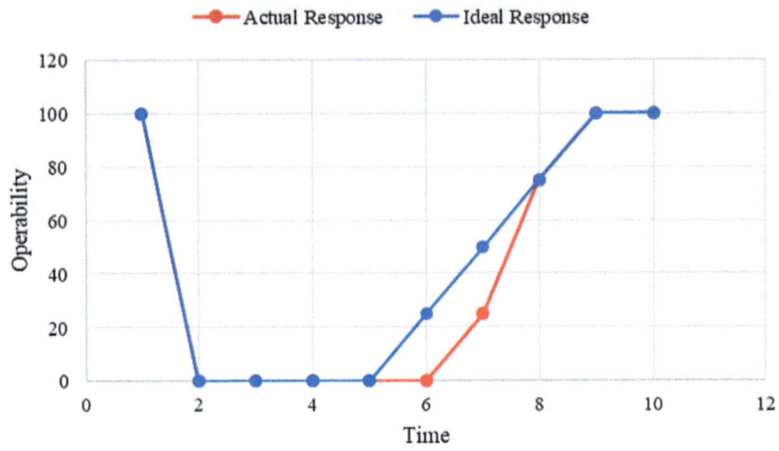

Fig. 9.3 The physical node response curve, if node (ij) was attacked at times 2 and 3. Blue represents the ideal reaction, and red represents the actual behavior enforced by the constraints

9.3.3.1 Preventing Multiple Attacks

Depending on the use case, it may not make sense for multiple attacks to take place on the same node. For example, if an attack exploits a specific vulnerability, the system owner could identify and mitigate this vulnerability once it has been identified. This can easily be incorporated by adding constraints such as

$$\sum_t z_{ijt} <= 1, \qquad \forall i, j.$$

More generally, consider that each node has a duration of time d_{ij} that must elapse before node (i, j) can be attacked again. If $d_{ij} = t_{ij}^{max}$, then the attack recovery has the property that a second attack cannot occur before there is full recovery from the initial attack. However, d_{ij} might be larger or smaller than t_{ij}^{max}. This type of constraint can be implemented by adding constraints:

$$\sum_{\tau=0}^{d_{ij}} z_{ij,t-\tau} \leq 1, \qquad \forall i, j, t.$$

When dealing with our specific family of response curves, if the required attack delay d_{ij} is greater than the node's linear recovery period $\Delta_{ij}^{(2)}$, the problem with multiple attacks in quick succession (outlined above) cannot occur. This condition is slightly less strict than the rule initially described where attacks cannot occur within a span of t_{ij}^{max}.

9.3.4 An Augmented Model for Operability Propagation if Responses to Attack Are Delayed

The following optimization model, viewed from the attacker's perspective, incorporates a delay in operability propagation (as derived in Sect. 9.2.2) and allows for time-varying physical node recovery.

Minimize

$$\sum_t w_t \phi_{*,t} \tag{9.25}$$

subject to

$$\phi_{i,t} \geq \gamma_{ji} P_{it}^j + \beta_{ji} - (1 - x_{ijt}^1)M, \qquad t = 0,\ \forall i, j, \tag{9.26}$$

$$\phi_{i,t} \geq \sum_{j=1}^{n_i} \alpha_{ji} P_{it}^j + \delta_i - (1 - x_{it}^2)M, \qquad t = 0,\ \forall i, \tag{9.27}$$

$$\phi_{i,t} \geq P_{\ell t}^0 - (1 - x_{it}^3)M, \qquad t = 0,\ \forall i,\ \forall \ell \in \mathcal{F}_i, \tag{9.28}$$

$$\phi_{i,t} \geq (1 - \rho_i)\phi_{i,t-1} + \rho_i \left(\gamma_{ji} P_{it}^j + \beta_{ji}\right) - (1 - x_{ijt}^1)M, \qquad \forall i, j, t > 0, \tag{9.29}$$

$$\phi_{i,t} \geq (1 - \rho_i)\phi_{i,t-1} + \rho_i \left(\sum_{j=1}^{n_i} \alpha_{ji} P_{it}^j + \delta_i\right) - (1 - x_{it}^2)M, \qquad \forall i, t > 0, \tag{9.30}$$

$$\phi_{i,t} \geq (1 - \rho_i)\phi_{i,t-1} + \rho_i \phi_{\ell,t} - (1 - x_{it}^3)M, \qquad \forall i, t > 0, \ell \in \mathcal{F}_i, \tag{9.31}$$

$$1 = \sum_{j=1}^{n_i} x_{ijt}^1 + x_{it}^2 + x_{it}^3, \qquad \forall i, t, \tag{9.32}$$

$$P_{it}^j \geq T_{ij} - \sum_{\tau \leq t} \left(T_{ij} - R_{ij}(t - \tau)\right) z_{ij\tau}, \qquad \forall (i, j), t \tag{9.33}$$

$$P_{it}^j \geq T_{ij} b_{ij}, \qquad \forall i, j, t, \tag{9.34}$$

$$B \geq \sum_{i,j,t} c_{ij} z_{ijt}, \tag{9.35}$$

$$z, X \text{ binary.} \tag{9.36}$$

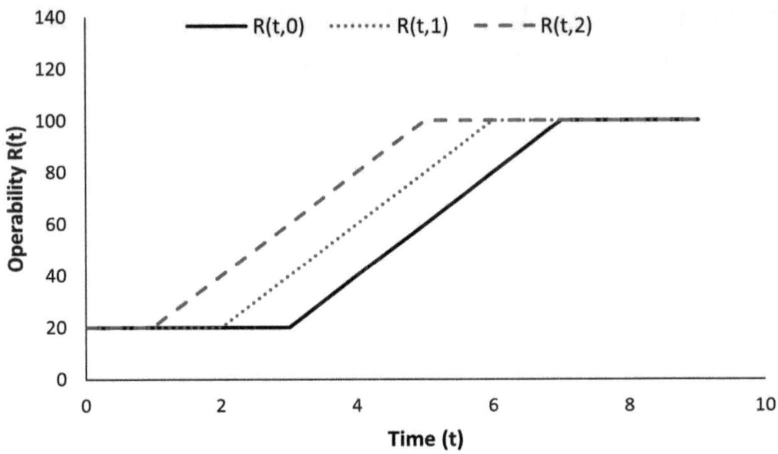

Fig. 9.4 The response curve for a physical node under three different investment strategies: $k = 0$, $k = 1$, and $k = 2$. Each unit invested reduces the node downtime by one period

9.4 A Defensive Perspective

The model formulations presented in the above are designed to determine the most devastating attacks that an attacker can carry out against the network. The results from such models can be helpful to understand which nodes are most critical to the network. Furthermore, this approach allows a defender to evaluate how certain defensive actions impact the network vulnerability and nodal criticality. In this section, we introduce a new class of defensive actions, where the physical node downtime can be reduced for a cost. We aim to use the previous optimization model as a way to assess this class of defensive actions and guide us to the best possible set of investments for a defender.

9.4.1 Defensive Decisions

Consider the option for a network owner to invest in the physical nodes in a manner that reduces their downtime after attack. More precisely, using the terminology from (9.23), the defender has an ability option to reduce the physical node parameter $\Delta_{ij}^{(1)}$ by k units for a cost for some node (i, j). The geometric interpretation behind such an investment is pictured in Fig. 9.4.

With this defensive option, the response function $R_{ij}(t)$ can be rewritten as a function of time after attack, as well as k, the amount invested. This is shown in Eq. (9.37).

$$R_{ij}(t-k) = \begin{cases} T_{ij}b_{ij}, & \text{for } t = 0, 1, \ldots, \Delta_{ij}^{(1)} - k, \\ T_{ij}b_{ij} + T_{ij}(1-b_{ij})\left\lceil \frac{t-\Delta_{ij}^{(1)}-k}{\Delta_{ij}^{(2)}} \right\rceil, & \text{for } t = \Delta_{ij}^{(1)} - k, \ldots, \Delta_{ij}^{(1)} + \Delta_{ij}^{(2)} - k, \\ T_{ij}, & \text{for } t \geq \Delta_{ij}^{(1)} + \Delta_{ij}^{(2)} - k. \end{cases} \tag{9.37}$$

To encode the defensive investment decisions, let y_{ijk} be a binary variable where $y_{ijk} = 1$ indicates the physical node (i, j) has its downtime reduced by k time units, and otherwise for $y_{ijk} = 0$, for $k = 0, \ldots, \Delta^{(1)}$. Using this new notation, we can rewrite constraints (9.24) to be

$$P_{it}^{j} \geq T_{ij} - \sum_{k=0}^{\Delta_{ij}^{(1)}} \sum_{\tau \leq t} \big(T_{ij} - R_{ij}(t-\tau-k)\big)\, z_{ij\tau}\, y_{ijk}, \qquad \forall (i,j),\ t \tag{9.38}$$

and include the additional condition:

$$\sum_{k=0}^{\Delta_{ij}^{(1)}} y_{ijk} = 1, \qquad \forall (i,j). \tag{9.39}$$

By virtue of constraint (9.39), the summand of (9.38) will be nonzero for at most one value of k (corresponding to the value of k for which $y_{ijk} = 1$). For this value of k, the constraint simplifies to

$$P_{it}^{j} \geq T_{ij} - \sum_{\tau \leq t} (T_{ij} - R_{ij}(t-\tau-k)) z_{ij\tau}, \tag{9.40}$$

which is the desired behavior. If $z_{ij\tau} = 0$ for all τ, then no attacks took place and $P_{it}^{j} = T_{ij}$. If instead, $z_{ij\tau} = 1$ for some τ, then this simplifies to $P_{it}^{j} \geq R_{ij}(t-\tau-k)$, as expected.

9.4.2 The Defensive Problem

Working with this class of defensive decisions, it is natural to ask where should investments be made to best fortify the network against possible attacks? In this section, we consider the bi-level optimization problem where the defender first chooses the nodes to invest in, subject to their defense budget, and the attacker then selects nodes to disable, subject to their attack budget. These sequential decisions can be considered a Stackelberg Game, where the defender is the leader and the attacker is the follower.

For any defensive investment decisions y and attack decisions z, let $U(y, z)$ equal the total value of the network objective (according to (9.4)). The bi-level optimization problem can then be stated as

$$\max_{y} \min_{z} U(y, z). \tag{9.41}$$

This formulation determines the investments y to minimize the attacker's ability to harm the network. The full optimization model with all associated constraints will be spelled out explicitly in Appendix section "Appendix A: Bi-level Optimization Model". As the attacker's decisions are made after those of the defender, the optimal z can be written as a function of y. We will refer to this function $A_F(y)$ as the follower's recourse:

$$z^* = A_F(y) = \arg\min_{z} U(y, z). \tag{9.42}$$

This recourse function is equivalent to solving the attacker's optimization problem from Sect. 9.3.4, including the fixed decisions y and constraints discussed in Sect. 9.4.1. Therefore, the bi-level optimization problem can be written as

$$\max_{y} U(y, A_F(y)). \tag{9.43}$$

In this way, the bi-level optimization problem reduces to maximizing a function over a single set of decisions y. However, this function in y must be treated essentially as a black box, which prevents many exact optimization techniques. Depending on the size of the network and the number of investment possibilities, it may be possible to do an exhaustive search over the decisions y. This would provide an exact solution, although it could be prohibitively time-intensive. Therefore, this approach is not recommended unless there are a small number of defensive courses of action that were decided upon beforehand.

There are many other heuristic approaches that can be used to find an acceptable, although not necessarily optimal, solution for y. Evolutionary approaches such as the genetic algorithm will start with a pool of candidate solutions and search for new candidates using the characteristics of those solutions with the best performance. Local search methods can try to iteratively make small, local improvements to find an optimum. Perhaps most simply, a subset of solutions that are sufficiently different (although not exhaustive) can be evaluated, and the best solution of that subset returned.

We propose an iterative algorithm that alternates between evaluating follower and leader actions. To do this, we define the *leader's recourse* function for a fixed z as

$$y^* = A_L(z, \mathcal{L}) = \arg\max_{y \notin \mathcal{L}} U(y, z). \tag{9.44}$$

For $\mathcal{L} = \emptyset$, this is equivalent to finding the best defensive investment y to prevent the specific attack z. This optimization problem is explicitly formulated in Sect. 9.6. Likewise, we slightly modify the definition of the follower's recourse to forbid certain attacks from recurring,

$$A_F(y, \mathcal{F}) = \arg\min_{z \notin \mathcal{F}} U(y, z). \tag{9.45}$$

Both functions are used in the following iterative algorithm. The idea is to start with a leader action y and find the best follower recourse $z = A_F(y, \emptyset)$. Once the leader decisions have been evaluated in this way, they are added to the set $\mathcal{L}$ to ensure the same decisions are not evaluated again. Then, to pivot to a new leader action, we find the optimal leader recourse $A_L(z, \mathcal{L})$, subject to the condition that it has not already been evaluated. This process then repeats, up to a maximum number of times (*itLim*) since the last improvement was seen. This iterative approach can be applied to a single starting solution or a set of possible starting solutions Λ. At all times in the algorithm, y_{best}, z_{best}, and U_{max} correspond to the best values of y, z, and U, respectively. The full algorithm is shown below:

Algorithm 10 An iterative algorithm to find the Stackelberg equilibrium

$\mathcal{L} = \emptyset$
$\mathcal{F} = \emptyset$
$U_{max} = 0$
for $y \in \Lambda$ **do**
    $i = 0$
    while $i < itLim$ **do**
        $z = A_F(y, \emptyset)$
        if $U(y, z) > U_{max}$ **then**
            $U_{max} = U(y, z)$
            $y_{best} = y$
            $z_{best} = z$
            $i = 0$
        else
            $i{+}{+}$
            $\mathcal{F} = \mathcal{F} \cup \{z\}$
            $F = A_F(y, \mathcal{F})$
        end if
        $\mathcal{L} = \mathcal{L} \cup \{y\}$
        $y = A_L(z, \mathcal{L})$
    end while
end for
return $y_{best}, z_{best}, U_{max}$

Note that this algorithm is another heuristic. It may perform well in practice, although there is no guarantee that it will find the optimal solution. In preliminary investigations, this approach settled on the optimal solution quickly, although our testing was limited. This method, along with other benchmarks, should be tested

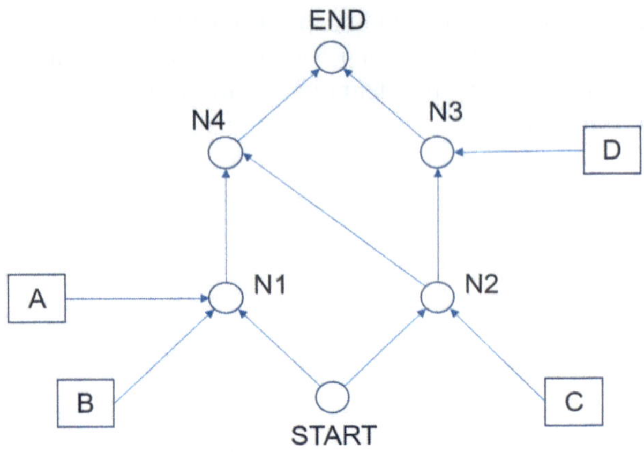

Fig. 9.5 A simple example of a hybrid mission network. Circular nodes are functional, and square nodes are physical

Table 9.1 Physical node parameters

Physical node	α_{ji}	β_{ji}
A	0.75	15
B	0.8	100
C	0.8	10
D	0.5	50

Table 9.2 Functional node parameters

Functional node	δ_i
N1	0
N2	20
N3	50
N4	0

against a more diverse set of test cases before a strong recommendation can be made regarding the most effective solution approach.

9.5 An Illustrative Example

In this section, we provide a simple network example to help illustrate how this model works. Consider the following network, pictured in Fig. 9.5. The principal network consists of six functional nodes, depicted as circles: *START*, $N1$, $N2$, $N3$, $N4$, and *END*. Attached to three of these functional nodes are physical nodes A, B, C, and D, which are depicted as squares.

Tables 9.1 and 9.2 show the nodal parameters α_{ji}, β_{ji}, and δ_i for this example. For simplicity, we define the parameter $\gamma_{ji} = 1$ for every physical node (i, j). Likewise, in this section, we will use $\rho_i = 1$ for every functional node i (i.e., the case of "instant dynamics"). For a more complete treatment of how ρ affects the solution, see Appendix section "Appendix C: The Impact of the Speed of Change of the Operabilities".

By substituting these parameters, the functional node operabilities are governed by the following equations.[1]

$$
\begin{aligned}
\phi_{START,t} &= 100 \\
\phi_{1t} &= \min\left[P_{1t}^{A} + 15,\ P_{1t}^{B} + 100,\ 0.75P_{1t}^{A} + 0.8P_{1t}^{B},\ \phi_{START,t}\right] \\
\phi_{2t} &= \min\left[P_{2t}^{C} + 10,\ 0.8P_{2t}^{C} + 20,\ \phi_{START,t}\right] \\
\phi_{3t} &= \min\left[P_{3t}^{D} + 50,\ 0.5P_{3t}^{D} + 50,\ \phi_{2t},\ 100\right] \\
\phi_{4t} &= \max\left[\phi_{1t},\ \phi_{2t}\right] \\
\phi_{END,t} &= \max\left[\phi_{3t},\ \phi_{4t}\right].
\end{aligned}
$$

From these equations some immediate observations can be made to help us understand the solution.

- The parameter $\beta_{B,1} = 100$ implies that $P_{1t}^{B} + 100$ will never be the unique minimum for determining the value of ϕ_{1t}. This second argument can effectively be removed from the ϕ_{1t} equation.
- ϕ_{1t} is more sensitive to changing values in A than changing values in node B. For example, setting $P_{1t}^{A} = 0$ and $P_{1t}^{B} = 100$ results in $\phi_{1t} = 15$, whereas setting $P_{it}^{A} = 100$ and $P_{1t}^{B} = 0$ only causes $\phi_{1t} = 75$. This suggests that it is more effective for an attacker to disable node A rather than node B.
- The second argument in the equation for ϕ_{3t} will always dominate the first. Therefore, the first argument, $P_{3t}^{D} + 50$, can effectively be removed.
- Nodes 4 and *END* have no associated physical nodes. Therefore, their operability relies solely on the path with the highest operability up to that point in the network.
- By substituting the equation for ϕ_{4t} into that for $\phi_{END,t}$, we see that the end node operability is actually the maximum of ϕ_{1t}, ϕ_{2t}, and ϕ_{3t}.

There is not a single common node (i.e., a bridge) among the three distinct paths for *START* to *END* ($N1 - N4$, $N2 - N4$, $N2 - N3$). This implies that disabling any single node will not affect the *END* node operability. Therefore, we allow an attacker budget of two nodes. We slightly weight the objective function to discount future time periods, making it more desirable for an attacker to disable nodes as early as possible in the time horizon.

[1] To simplify notation, we drop the preceding N from functional node IDs. For example, node $N1$ simply becomes node 1.

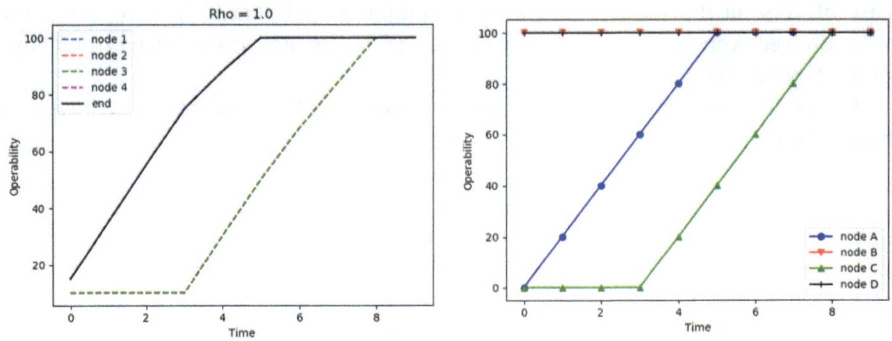

Fig. 9.6 The operability curves showing the Stackelberg solution for the simple example from Fig. 9.5. The graph on the left shows the operability of the functional nodes, and the graph on the right shows the operability of the physical nodes

The last remaining data that is needed is the parameters governing the physical node downtime. We assume that each node, when attacked, is down for three time periods and undergoes a linear recovery for the following five time periods. The defender can choose to reduce the downtime of a physical node by one time period, a maximum of three times total (not necessarily all on the same node). With the notations introduced earlier, $\Delta^{(1)} = 3$, $\Delta^{(2)} = 5$, and $B = 3$.

The charts below show the solution to the bi-level optimization problem over ten time periods, assuming all physical nodes are initially fully operable, i.e., set to 100. The optimal leader (Defense) strategy is to invest all three units of defense into reducing the downtime for physical node A. The optimal follower (Attack) strategy is to immediately disable nodes A and C. These are the two lines below 100 shown in the right graph in Fig. 9.6. Since the downtime of node A was reduced to zero, it starts the linear recovery phase from time zero. While node C is disabled and recovering, the maximum operability in the network passes through nodes $N1-N4$, which is what is seen in the left graph in Fig. 9.6. The black line for the *END* node operability completely overlaps those for nodes $N1$ and $N4$, and the green line for the operability of nodes $N2$ and $N3$ completely overlaps.

This solution is intuitive because nodes A and C are tied to functional nodes $N1$ and $N2$, respectively, which cover every *START* – *END* path. Since the network operability derives from the maximum operability of the *START* – *END* paths, the defender only needs to invest in reducing the downtime of any single path. That is, it does not make sense to invest in both A and C, because that would just take longer for the path through $N1$ to regain its maximum operability, which is all that matters for this network. A very similar solution could be achieved by investing all units in node C, but it would be just slightly worse, because ϕ_{2t} is bounded below by 10 as opposed to ϕ_{1t}, which is bounded below by 15. In our case, this implies that the operability along path $N1 - N4$ would slightly exceed that along $N2 - N3$ or $N2 - N4$, and therefore it makes more sense for the defender to invest in this path.

9.6 Conclusion

This chapter extended the static analyses of the preceding chapters by introducing temporal dynamics into the analysis of hybrid mission network. We modeled (i) detection delays for disruptions to physical nodes, (ii) recovery dynamics, and (iii) time-dependent node operability. We examined these dynamics from both sides of the engagement—first with an attacker who ignores defender adaptation and then with a defender who proactively fortifies while anticipating attacker adaptation. A worked example shows how detection delay, recovery time, and response policy jointly shape mission performance over time.

Appendix A: Bi-level Optimization Model

This appendix shows the full bi-level optimization model, each piece of which has separately been discussed in Eq. (9.41). Solving this optimization problem gives the Stackelberg Equilibrium, which is the set of investments a leader/defender should make to minimize the potential damage that an adversary can do to their network. This full model was denoted $U(y, A_F(y))$ in Eqs. (9.42) and (9.43).

$$\underset{y}{\text{Max}}\ \underset{z,X}{\text{Min}}\text{:} \sum_t w_t \phi_{*,t} \tag{9.46}$$

subject to

$$\phi_{i,t} \geq \gamma_{ji} P_{it}^j + \beta_{ji} - (1 - x_{ijt}^1)M, \qquad t = 0,\ \forall i, j, \tag{9.47}$$

$$\phi_{i,t} \geq \sum_{j=1}^{n_i} \alpha_{ji} P_{it}^j + \delta_i - (1 - x_{it}^2)M, \qquad t = 0,\ \forall i, \tag{9.48}$$

$$\phi_{i,t} \geq P_{\ell t}^0 - (1 - x_{it}^3)M, \qquad t = 0,\ \forall i,\ \forall \ell \in \mathcal{F}_i \tag{9.49}$$

$$\phi_{i,t} \geq (1 - \rho_i)\phi_{i,t-1} + \rho_i \left(\gamma_{ji} P_{it}^j + \beta_{ji}\right) - (1 - x_{ijt}^1)M, \qquad \forall i, j, t > 0, \tag{9.50}$$

$$\phi_{i,t} \geq (1 - \rho_i)\phi_{i,t-1} + \rho_i \left(\sum_{j=1}^{n_i} \alpha_{ji} P_{it}^j + \delta_i\right) - (1 - x_{it}^2), \qquad \forall i, t > 0, \tag{9.51}$$

$$\phi_{i,t} \geq (1 - \rho_i)\phi_{i,t-1} + \rho_i \phi_{\ell,t} - (1 - x_{it}^3)M, \qquad \forall i, t > 0, \ell \in \mathcal{F}_i, \tag{9.52}$$

$$1 = \sum_{j=1}^{n_i} x_{ijt}^1 + x_{it}^2 + x_{it}^3, \qquad \forall i, t, \tag{9.53}$$

$$P_{it}^j \geq T_{ij} - \sum_{k=0}^{\Delta^{(1)}} \sum_{\tau \leq t} \left(T_{ij} - R_{ij}(t - \tau - k)\right) z_{ij\tau} y_{ijk}, \forall (i, j), t, \tag{9.54}$$

$$P_{it}^j \geq T_{ij} b_{ij}, \qquad \forall i, j, t, \tag{9.55}$$

$$B_F \geq \sum_{i,j,t} c_{ij}^F z_{ijt}, \tag{9.56}$$

$$B_L \geq \sum_{i,j,t} c_{ij}^L y_{ijk}, \tag{9.57}$$

$$1 = \sum_{k=0}^{\Delta^{(1)}} y_{ijk}, \qquad \forall i, j, \tag{9.58}$$

$$z, x \text{ binary}. \tag{9.59}$$

Appendix B: Leader's Recourse

The optimization model below shows the leader's recourse. That is, it assumes the attack values z_{ijt} are fixed, and aims to choose appropriate defensive investments y_{ijk}, to shorten the time $\Delta_{ij}^{(1)}$ to begin to respond to the disabling attack of physical node (i, j), to best compensate against these attacks. In Sect. 9.4.2, this model was denoted as $A_L(z, \emptyset)$.

Maximize

$$\sum_t w_t \phi_{*,t} \tag{9.60}$$

subject to

$$\phi_{i,t} \leq (1 - \rho_i)\phi_{i,t-1} + \rho_i(\gamma_{ji} P_{it}^j + \beta_{ji}), \qquad \forall i, j, t, \tag{9.61}$$

$$\phi_{i,t} \leq (1 - \rho_i)\phi_{i,t-1} + \rho_i \left(\sum_{j=1}^{n_i} \alpha_{ji} P_{it}^j + \delta_i \right), \qquad \forall i, t, \tag{9.62}$$

$$\phi_{i,t} \leq (1 - \rho_i)\phi_{i,t-1} + \rho_i \phi_{\ell t} + (1 - x_{i\ell t})M, \qquad \forall i, t\ \forall \ell \in \mathcal{F}_i, \tag{9.63}$$

$$1 = \sum_{\ell \in \mathcal{F}_i} x_{i\ell t}, \qquad \forall i, t, \tag{9.64}$$

$$P_{it}^{j} \leq T_{ij} - \sum_{k=0}^{\Delta^{(1)}} \sum_{\tau=t-\Delta^{(1)}-\Delta^{(2)}}^{t} (T_{ij} - R_{ij}(t-\tau-k)) z_{ij\tau} y_{ijk}, \qquad \forall\, i, j, t, \tag{9.65}$$

$$1 = \sum_{k=0}^{\Delta^{(1)}} y_{ijk}, \qquad \forall\, i, j, \tag{9.66}$$

$$B_L \geq \sum_{i,j,k} c_{ijk} y_{ijk}, \tag{9.67}$$

$$x, y \text{ binary}. \tag{9.68}$$

This is a relatively straightforward analog of the attacker model, $A_L(z, \emptyset)$. The primary difference is that from a defensive point of view, the network operability should be maximized instead of being minimized. Therefore, the directions of the inequalities (9.61)–(9.63) and (9.65) are switched compared to the original attacker model. All other constraints have been explained in earlier portions of the paper.

Defensive Constraints

In the leader's recourse formulation above, constraint (9.65) is slightly altered. As each attack z lowers the upper bound on P_{it}^{j}, the operability of physical node j that supports functional node i, it is possible that multiple attacks in close succession can "constructively interfere" and cause an infeasibility. By "constructively interfere" we mean that if for some t and all k, one attack begins before the impact of the previous attack ends, i.e.,

$$\sum_{\tau=t-\Delta^{(1)}-\Delta^{(2)}}^{t} (T_{ij} - R_{ij}(t-\tau-k)) z_{ij\tau} > T_{ij} - T_{ij} b_{ij}. \tag{9.69}$$

In this case, from the defensive constraint (9.65) we get the first inequality in

$$P_{it}^{j} \leq T_{ij} - \sum_{\tau=t-\Delta^{(1)}-\Delta^{(2)}}^{t} (T_{ij} - R_{ij}(t-\tau-k)) z_{ij\tau} < T_{ij} b_{ij} \tag{9.70}$$

for whichever value of k that satisfies $y_{ijk} = 1$. The second inequality follows from (9.69). However, (9.70) is infeasible because variable P_{it}^{j} is also bounded below by $T_{ij} b_{ij}$ as illustrated in Fig. 9.1.

To correct this issue, we can ensure that the total summand value never exceeds $T_{ij} - T_{ij} b_{ij}$. The reformulated constraint is shown below.

$$P_{it}^{j} \leq T_{ij} - \sum_{k=0}^{\Delta^{(1)}} \min\left(T_{ij} - T_{ij}b_{ij}, \sum_{\tau=t-\Delta^{(1)}-\Delta^{(2)}}^{t} (T_{ij} - R_{ij}(t-\tau-k))z_{ij\tau} \right) Y_{ijk}$$

$$\forall\, i, j, t. \tag{9.71}$$

This will ensure that attacks that are close together can never inadvertently cause an infeasibility. Since the values of z_{ijt} are known for the leader's recourse, rewriting the constraint with this maximum only affects how the coefficients of the IP are calculated; it does not introduce any nonlinearity to the problem.

Note that this minor detail only exists in the leader's recourse problem, because of the direction of the inequality in the constraint that links z_{ijt} and y_{ijk} to P_{it}^{j}. Since the follower's recourse is a minimization, the inequality for the corresponding constraint is reversed, which prevents any possibility of infeasibility.

Appendix C: The Impact of the Speed of Change of the Operabilities

The parameter ρ_i relates to the speed at which functional node i operability can change. As such, ρ_i limits how quickly the functional node operability is affected by the changes in its associated physical nodes. Likewise, ρ_i also impacts how long a functional node's operability will take to return to steady state conditions. Setting $\rho_i = 1$ indicates that these changes in operability are instantaneous. On the other end of the spectrum, setting $\rho_i = 0$ effectively prevents any operability from propagating through the functional nodes. That is, physical node impact will not be felt and operability will no longer "flow" along the paths in the network.

In Sect. 9.5, ρ_i was set to a constant value of ρ for all functional nodes. The graphs in Fig. 9.7 illustrate how reducing the value of ρ equally across all nodes impacts the operability curves. The curves shown all correspond to the Stackelberg Equilibrium, which did not change regardless of the value of ρ. The same attacker and defender strategies as described in Sect. 9.5 are still optimal.

If $\rho < 1$, the operability curve will approach but not reach its equilibrium value in finite time. However, as ρ decreases, the rate of convergence to the equilibrium decreases and the operability curves tend to be much smoother.

Table 9.3 shows the runtimes for calculating the Stackelberg Equilibrium via exhaustive search for several different values of ρ. As ρ decreases from 1 to 0.5, the time to solve these problems increases. With lower values of ρ the problem gets more complex because any action affects many subsequent time periods. However, for values of ρ below 0.5, we see that the time starts to decrease again. One possible explanation is that for a small ρ, the changes in subsequent time periods become less significant, so the functional node behavior is basically only determined by the initial actions at $t = 0$. This explanation is just a hypothesis, and

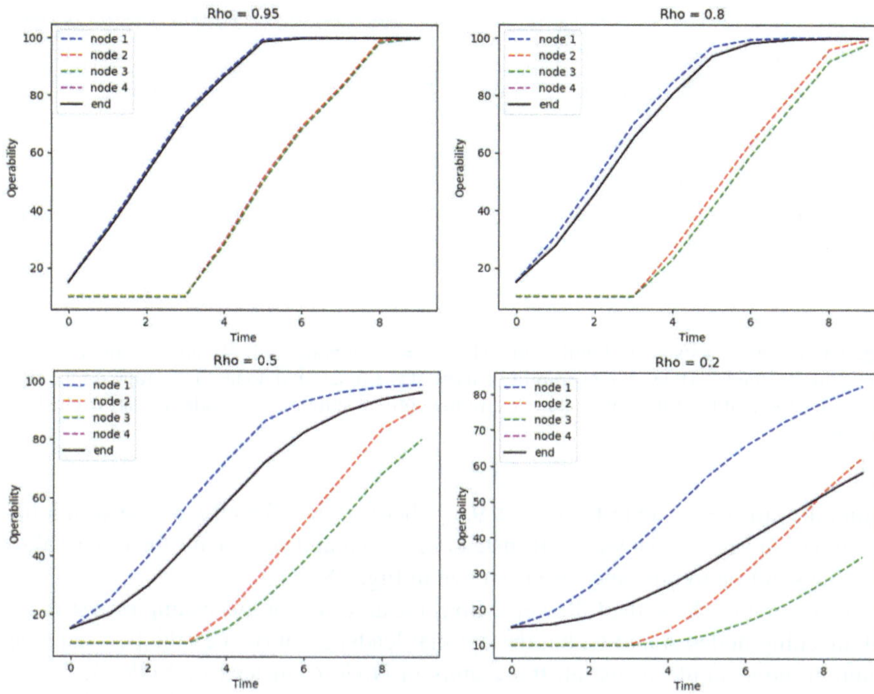

Fig. 9.7 Plots of the resulting functional node operability values, for $\rho \in \{0.95.0.8, 0.5, 0.2\}$

Table 9.3 Run times for finding Stackelberg equilibrium in our example network, dependent on global parameter of ρ. Times are based on performing an exhaustive search of the leader actions

ρ	1.0	0.95	0.9	0.8	0.7	0.6	0.5	0.4	0.3	0.2	0.1
Run time (s)	4.6	10.3	15.5	24.1	22.3	26.4	30.4	16.5	5.6	3.5	3.2

more experimentation would need to be done to verify that this relationship holds consistently.

One interesting feature of this problem is that the solution (i.e., the Leader invests 3 in A, and the Follower attacks A and C at time 0) was invariant regardless of ρ. This observation could lead to a simple approximation, where we assume that $\rho = 1$, and the solution to this first problem can then be used to directly compute the functional node operabilities with a new value of $\rho \neq 1$. This has the benefit of solving an optimization problem that has been much quicker to solve and easier to interpret, while still generating a solution that follows the ρ relationships for your use case. Again, this method is not guaranteed to work, but it may merit further investigation.

All of the examples shown thus far have assumed that a constant value of ρ applies to all functional nodes equally within the network. When this is not the case, the values of ρ_i can also impact the Stackelberg Equilibrium. Consider, once

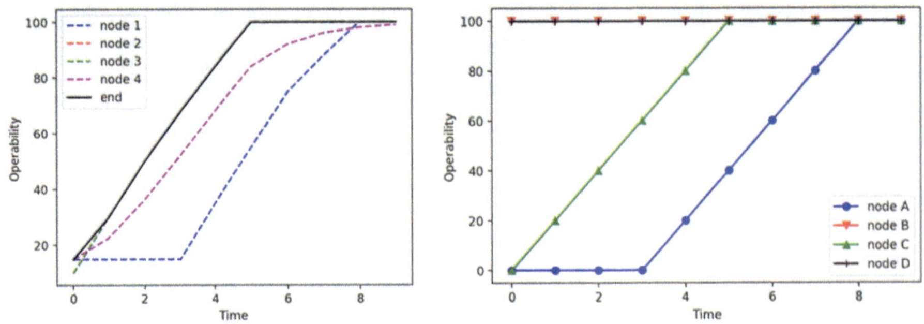

Fig. 9.8 The new Stackelberg Equilibrium when $\rho_i = 1$ for nodes $N1$, N_2, and N_3, and $\rho_4 = 0.5$. The graph on the left shows the operability curves of the functional nodes. The graph on the right shows the operability of the physical nodes. In the optimal solution, the leader invests exclusively in node C

again, the simple example from Sect. 9.5, where $\rho_i = 1$ for all i whose solution is illustrated in Fig. 9.6. However, if instead $\rho_4 = 0.5$ and $\rho_i = 1$ for the other values of i, the solution to this new case is shown in Fig. 9.8.

As the graphs show, the follower, who is the attacker for this example, is still best off attacking nodes A and C, but the optimal leader solution, i.e., the best defender solution, now should invest all three units in Node C instead of Node A. This is because it will now take longer for increases in operability (either from $N1$ or $N2$) to "register" with $N4$. This is seen in the left graph in Fig. 9.8, where the purple dashed curve for $N4$ is slightly below that of its predecessor $N2$ (red dashed curve mostly covered by black). This slowness to adapt strongly devalues the paths that go through $N4$. Thus, the leader wants to bolster the remaining path $N2 - N3$, which is done by completely investing in node C.

Appendix D: Response Approximation Error Bounds

In this section, we compute the maximum error in the response function that can occur from using our simplified constraints (9.24). We deal only with the case of two consecutive attacks, understanding that attacking the same node more than twice in a short period of time is not likely to happen. However, the logic in this section could be extended to compute error bounds when more than two attacks occur.

Assume that two attacks take place on Node (i, j) so

$$P_{it}^{j} \geq T_{ij} - \left[(T_{ij} - R_{ij}(t - t_1)) + (T_{ij} - R_{ij}(t - t_2))\right]$$

We assume that both $(t - t_1)$ and $(t - t_2)$ are less than $\Delta^{(1)} + \Delta^{(2)}$, otherwise the quantity $T_{ij} - R_{ij}(t - t_k)$ vanishes for $k = 1$ or $k = 2$. Without loss of generality, we assume that $t_1 > t_2$.

First, consider the case where $t_1 - t_2 \geq \Delta^{(2)}$. The statements below show that $t - t_1 \leq \Delta^{(1)}$.

$$t \leq t_2 + \Delta^{(1)} + \Delta^{(2)} \qquad \text{(Only considering } t \text{ that satisfy this condition)} \tag{9.72}$$

$$t \leq (t_1 - \Delta^{(2)}) + \Delta^{(1)} + \Delta^{(2)} \qquad \text{(Per assumptions on this case)} \tag{9.73}$$

$$t \leq t_1 + \Delta^{(1)} \qquad \text{(simplifying)} \tag{9.74}$$

$$t - t_1 \leq \Delta^{(1)} \qquad \text{(rearranging)} \tag{9.75}$$

The implication is that $R_{ij}(t - t_1) = T_{ij}b_{ij}$, by definition. Since this is the lower-bound for P_{it}^j, then P_{it}^j will equal $T_{ij}b_{ij}$, regardless of the value of $R_{ij}(t - t_2)$. Therefore, in this case, the simplified constraint describes the ideal behavior.

Lastly, consider the case when $t_1 - t_2 \leq \Delta^{(2)}$. If $t - t_1 \leq \Delta^{(1)}$, the same conclusion holds as the previous case. Therefore, the only time periods t where the approximate constraints differ from the ideal behavior is when $t_1 + \Delta^{(1)} < t < t_2 + \Delta^{(1)} + \Delta^{(2)}$. This can be rewritten as

$$t \in (t_1', \ t_1' + \Delta^{(2)} - \delta)$$

where $t_1' = t_1 + \Delta^{(1)}$ and $\delta = t_1 - t_2$. Equivalently, $t_1' + 1 \leq t \leq t_1' + \Delta^{(2)} - \delta - 1$. This interval consists of $\Delta^{(2)} - \delta - 1$ discrete time values where the approximate constraints will be imperfect. Thus, the closer in time the two consecutive attacks occur, the more time periods will be affected.

To explicitly compute the error, we can use (9.24) where $z_{i,j,t_1} = z_{i,j,t_2} = 1$, and $z_{i,j,t} = 0$ for all remaining times t. Taking this definition in conjunction with the bound that $P_{it}^j \geq T_{ij}b_{ij}$, the value of P_{it}^j can be computed as

$$\begin{aligned}
P_{it}^j &= \max\left[T_{ij}b_{ij}, \ R(t - t_1) - (T_{ij} - R(t - t_2))\right] \\
&= \max\left[R(t - t_1) - (R(t - t_1) - T_{ij}b_{ij}), \ R(t - t_1) - (T_{ij} - R(t - t_2))\right] \\
&= \underbrace{R(t - t_1)}_{\text{Ideal response}} - \underbrace{\min\left[R(t - t_1) - T_{ij}b_{ij}, \ T_{ij} - R(t - t_2)\right]}_{\text{Error}}
\end{aligned}$$

If we assume that node operability should only be determined by the most recent attack (at t_1), we can see that the error in P_{it}^j is given by

$$\text{Error}(t) = \min\left[R(t - t_1) - T_{ij}b_{ij}, \ T_{ij} - R(t - t_2)\right]$$

In the range of interest, we parameterize t such that

$$t = t_1 + \Delta^{(1)} + k = t_2 + \delta + \Delta^{(2)} + k$$

for $k \in \{1, \dots, \Delta^{(2)} - \delta - 1\}$. Substituting these equations into the error function, and using the definition of our response function in Sect. 9.3.2, yields

$$\text{Error}(k) = \min\left[T_{ij}(1-b_{ij})\frac{k}{\Delta^{(2)}},\ T_{ij}(1-b_{ij})\frac{\Delta^{(2)}-\delta-k}{\Delta^{(2)}}\right].$$

As k increases, the first argument increases and the second decreases. They become equal when $k = (\Delta^{(2)} - \delta)/2$, which is where the error is maximized. Therefore, letting $m = \lceil \frac{\Delta^{(2)}-\delta}{2} \rceil$, we can further simplify:

$$\text{Error}(k) = \begin{cases} T_{ij}(1-b_{ij})\frac{k}{\Delta^{(2)}}, & k = 1, \dots, m-1, \\ T_{ij}(1-b_{ij})\frac{\Delta^{(2)}-\delta-k}{\Delta^{(2)}} & k = m, \dots, \Delta^{(2)} - \delta - 1. \end{cases}$$

By summing over all k, we can obtain the total error:

$$\begin{aligned} \text{Total Error} &= \frac{T_{ij}(1-b_{ij})}{\Delta^{(2)}}\left[\sum_{k=1}^{m-1} k + \sum_{k=m}^{\Delta^{(2)}-\delta-1} (\Delta^{(2)} - \delta - k)\right] \\ &= \frac{T_{ij}(1-b_{ij})}{\Delta^{(2)}}\left[\sum_{k=1}^{m-1} k + \sum_{j=1}^{\Delta^{(2)}-\delta-m} j\right] \\ &= \frac{T_{ij}(1-b_{ij})}{2\Delta^{(2)}}\left[m(m-1) + (\Delta^{(2)} - \delta - m)(\Delta^{(2)} - \delta - m + 1)\right]. \end{aligned}$$

To simplify this expression, we substitute back for m, depending on whether or not $\Delta^{(2)} - \delta$ is even or odd.

If $\Delta^{(2)} - \delta$ is even, then $m = (\Delta^{(2)} - \delta)/2$. In this case:

$$\begin{aligned} \text{Total Error} = \frac{T_{ij}(1-b_{ij})}{2\Delta^{(2)}}&\left[\frac{\Delta^{(2)}-\delta}{2}\left(\frac{\Delta^{(2)}-\delta-2}{2}\right) + \left(\Delta^{(2)} - \delta - \frac{\Delta^{(2)}-\delta}{2}\right)\right. \\ &\left.\times\left(\Delta^{(2)} - \delta - \frac{\Delta^{(2)}-\delta+2}{2}\right)\right] \\ = \frac{T_{ij}(1-b_{ij})(\Delta^{(2)}-\delta)}{4\Delta^{(2)}}&\left[\left(\frac{\Delta^{(2)}-\delta-2}{2}\right) + \left(\frac{\Delta^{(2)}-\delta+2}{2}\right)\right] \\ = \frac{1}{4\Delta^{(2)}}&T_{ij}(1-b_{ij})(\Delta^{(2)}-\delta)^2. \end{aligned}$$

If instead $\Delta^{(2)} - \delta$ is odd, then $m = (\Delta^{(2)} - \delta + 1)/2$. In this case:

$$
\begin{aligned}
\text{Total Error} &= \frac{T_{ij}(1-b_{ij})}{2\Delta^{(2)}}\left[\frac{\Delta^{(2)}-\delta+1}{2}\left(\frac{\Delta^{(2)}-\delta-1}{2}\right)\right.\\
&\quad \left.+\left(\Delta^{(2)}-\delta-\frac{\Delta^{(2)}-\delta+1}{2}\right)\left(\Delta^{(2)}-\delta-\frac{\Delta^{(2)}-\delta+1}{2}+1\right)\right]\\
&= \frac{T_{ij}(1-b_{ij})(\Delta^{(2)}-\delta-1)}{4\Delta^{(2)}}\left[\left(\frac{\Delta^{(2)}-\delta+1}{2}\right)+\left(\frac{\Delta^{(2)}-\delta+1}{2}\right)\right]\\
&= \frac{1}{4\Delta^{(2)}}T_{ij}(1-b_{ij})(\Delta^{(2)}-\delta-1)(\Delta^{(2)}-\delta+1).
\end{aligned}
$$

References

Cummings R (1919) The Girl in the Golden Atom. All-Story Weekly, New York

Garvey PR, Pinto CA, Santos JR (2014) Modelling and measuring the operability of interdependent systems and systems of systems: advances in methods and applications. Int J Syst Syst Eng 5(1):1–24

Guariniello C, DeLaurentis D (2017) Supporting design via the system operational dependency analysis methodology. Res Eng Des 28(1):53–69

Glossary

Antifragility A property of a system which paradoxically performs better as a result of previous shocks, stressors, or attacks to it. xii, xxiii, xxxv, 181, 184, 185

Atom Mapping The process of describing a system in terms of its most elementary components. 33

Attack path A sequence of nodes which an attacker transverses to attack a targeted node. 123–125

Bi-level optimization A type of optimization where the solution of one or more optimizations is part of the input to another of another optimization. 79, 187, 201, 202, 206, 207

Birnbaum cutset A heuristically derived approximate minimum cutset identified using Birnbaum Structural Importance measures. 108, 110, 111

Birnbaum Structural Importance A measure in reliability engineering that quantifies the importance of a component within a system based on the expected increase in the probability of mission failure when an individual system fails averaged over uncertainty of all other components. xxi, 53, 77, 93, 99, 104, 113, 115

Birnbaum Structural Index Nearly synonymous with the Birnbaum Structural Importance, but with an emphasis on its role as a metric. 101

Chebyshev's Inequality A statistical theorem that provides a simple upper bound on the probability that the value of a random variable deviates from its mean by more than a specified number of standard deviations. 105

Child contribution function The function that characterizes the relationship between a parent node and its dependent children nodes. This might be a single child contribution function, which expresses a dependency only on a single child node, or a multiple child contribution function, which expresses a dependency of two or more child nodes. 49–51

L. Servi, *Mission Dependency Network Analysis*, Risk, Systems and Decisions,
https://doi.org/10.1007/978-3-032-11260-6

Child node If node A of a mission dependent network depends on node B, i.e., the operability of node B impacts the operability of node A, node B is said to be the child node of node A. 5

Composite nodes Nodes with the property of both functional and physical nodes. 94

Continuous dependency representation A dependency representation where its operability parameter values are continuous-valued. 45

Critical path vector With respect to node i a critical path vector is the binary vector indicating which nodes in a mission are operable or not with the property that the mission will be operable if and only if node i is also operable. These vectors are used in the computation of the Birnbaum Structural Index. 99, 104–106, 110, 113

Criticality of Dependency One of the two parameters used in Functional Dependency Network Analysis (FDNA) to capture the dependency between a node and its dependent child node. This parameter sets an upper bound on the difference between the operability of a node and that of its dependent child node. This is sometimes abbreviated by COD. 5, 29, 30, 51, 59

Crown Jewel Analysis An approach to categorize with discrete-valued metrics the criticality of assets that are most critical to the accomplishment of a mission. 25, 49, 163

Cutset A set of nodes in a graph with the property that all paths from the Start node to the End node must pass through at least one node of this set. As such, if all nodes in a cutset are disabled, then the mission is disabled. xxii, 53, 104, 107, 108, 129

Degradation Index A measure of the criticality of a node found by (1) computing the change of the mission's operability as a result of the operability of a specific node changing by a very small amount, (2) calculating the ratio of these two changes, and then (3) taking the limit as the latter change approaches zero. 4–7, 10, 12, 15

Degree of node centrality The number of edges a node has incident to it. 77

Dependency representation The formal mathematical representation of the dependency between the nodes in a mission dependency network. 4, 47–52, 54, 58, 69, 136, 149, 163, 169, 179, 187

Depth-first search An algorithmic approach for traversing or searching tree or graph data structures by exploring as far down a branch as possible before backtracking to explore other branches. 101

Design Structure Matrix A compact matrix-based representation of a mission that captures the dependences among its components. The (i, j) element of such matrices conveys the parameters characterizing the dependencies between node i and node j. 50

Digital twin A dynamic, virtual representation of a specific physical asset, process, or system that stays synchronized with its real counterpart via data, enabling monitoring, simulation, and prediction across the system's life cycle. viii, 69

Directed acyclic graph A directed graph in which no cycles exist. 80

Dual The dual of a linear program is another linear program derived from the original (primal) linear program whose optimal objective has the same objective function and whose variables represent the marginal impact of each constraint in the primal problem on the optimal objective. 7, 9, 10, 13, 18–21, 131, 140, 143, 174

Fault Tree A graphical representation used to analyze the nodes whose failure can cause mission failures by mapping out the various logical relationships between potential faults and their effects. 49–53, 59, 61

Follower's recourse In a two-person game with a leader and a follower, this is the best action the follower should make given knowledge of the leader's action. The leader selects their best action by optimizing their objective while using knowledge of the follower's recourse.. 202, 203, 210

Functional Dependency Network Analysis (FDNA) A two-parameter framework to quantify the dependency relationship between pairs of nodes in a mission. xix, 3, 25, 49, 78, 118, 136, 162

Functional network A network that consists of functional nodes rather than physical nodes. 119, 122, 123

Functional node Functional nodes represent the capability to perform some action. The operability of a functional node might depend on the operability of certain subsets of physical nodes, called supporting combinations, and other functional nodes, if any, they depend on. See Hybrid Networks. 78, 80, 86, 94, 98, 101–103, 117–119, 121, 122, 124, 126, 132

Fussell-Vesely importance The fraction of mission failure probability attributable to a specific component of the mission. 77, 93

Game theory Mathematical models of strategic interactions among rational decision-makers where the outcome for each decision-maker depends on the choices of all other decision-makers. In some cases, the decision-makers are assumed to make their decisions at the same time. In other cases, the game specifies the order of the decision-maker's moves. 161, 162, 185

Genetic algorithm A search heuristic inspired by the process of natural selection, used to solve optimization and search problems by iteratively evolving a population of candidate solutions through selection, crossover, and mutation. 53, 54, 62–64, 202

Global clustering coefficient A measure of the overall tendency of nodes in a network to cluster together, calculated as the ratio of the number of closed triplets, a set of three nodes which are connected to each other, to the total number of triplets, a set of three nodes where at least two of them are connected and closed, in the network. 53, 93

Greedy algorithm An approximate optimization algorithmic approach that makes a series of choices, with the goal of finding a locally optimal solution at each step. The myopic nature of the algorithm creates the potential for substantially faster time to compute a solution but typically does not find the best solution. 53, 54, 65, 66, 68

Ground Operability A parameter of the System Operational Dependency Analysis (SODA) which captures the operability of a non-leaf node before the influence of the operability of its dependent child nodes is accounted for. 32

Group Criticality A measure of the extent that a group of nodes within a system together can impact the functionality of the system. 32

Hierarchical network A network whose topology has layers. As such, nodes at layer n may only be connected to nodes at layer $n + 1$ or nodes at layer $n - 1$. 22

Hybrid mission network A mission network that consists of both physical and functional nodes. xxxiii, xxxv, 78–80, 86, 88–90, 94, 187, 188, 204, 207

Hybrid network representation A network representation that consists of nodes which are either physical or functional nodes or both. 94, 102

Impact of Dependency A parameter of the System Operational Dependency Analysis (SODA) which is typically very close to 0 and guarantees that a parent node's operability will be 0 if all of its children nodes' operability is 0. 32

In-degree The number of edges going into a node. 88

Large Language Model Very large, typically transformer-based neural networks trained to predict the next token in text. By learning this distribution from massive corpora, they can generate and manipulate language and perform many tasks via prompts.. ix

Leader's recourse In a two-person game with a leader and a follower, this is the best action the leader should make given knowledge of the follower's best action. 202, 209, 210

Leaf node A node in a dependency network whose operability does not depend on the operability of other nodes. 5, 7, 34, 38, 39, 41, 43, 51, 65–67, 118–126, 136, 149, 151, 152, 166–168

Linear programs A class of optimization models used extensively in Operations Research having an objective function with linear coefficients and either equality or inequality constraints which all have linear coefficients. xi, xix, xxiii

MAX The maximization function. 28

McCormick Envelope A technique used in optimization to approximate the feasible region of a bilinear function by constructing convex and concave envelopes that bound the bilinear term. 144

Measure of Effectiveness A metric capturing the ability to attain a desired end state, an achievement of an objective, or the creation of an effect. 117, 118, 136, 162, 184

Measure of Performance A metric quantifying the success of a mission. 151

Minimum cutset The smallest set of nodes with the property that all paths from the start of the network to the end of the network go through at least one node in this set. As such, if all such nodes are disabled, then the mission is disabled. 107, 108, 110, 111

Mission dependency network A network of nodes typically modeling a mission for which each link corresponds to a dependency relationship. x, xix, xx, xxii, xxiii

Mission network A network of functional nodes whose topology characterizes activities which must be successfully taken, in a certain order for the mission to be successful. x, xx–xxiii, xxxix, 4, 6, 10, 70, 78, 80, 117, 118, 135, 136, 138, 161–164, 169, 170, 175, 179, 182, 184, 187

Mission Network Pivot Algorithm An optimization technique used in the network analysis to dynamically adjust and reconfigure mission-critical networks, ensuring optimal performance and resilience by pivoting to alternative network paths or configurations in response to changes or disruptions. 174–177, 179

Mixed Integer Linear Program A linear optimization program where some variables are constrained to be real numbers and others integers. 26, 162, 170, 196

Multi-child contribution One of the two qualitatively different contributions that capture the dependency of one node on other node(s). The multi-child contribution captures how all of the dependent nodes impact a single node. For example, a multi-child contribution might be a weighted average of all its dependent nodes' metric value. The other type of contribution is a single child contribution. 56, 57

Multi-commodity single path formulation An optimization approach in network flow problems where multiple commodities are routed through a network, with each commodity requiring a single, dedicated path from its source to its destination, while optimizing overall network performance or cost. 124, 125

Nash Equilibrium The decisions of a multi-person game with the property that no player is better off changing their decision if their opponent's decision remains unchanged. 181

Node metric This is an indicator of the health or functionality of a node. 50, 55, 57, 62, 65

Nominal Optimal Solution The solution of a optimization program without added constraints. 169, 177

Operability value The quantitative measure of the functionality of a system or a mission. 4, 47, 48, 51, 52, 56, 57, 59, 68, 69, 149, 151

Out-degree The number of edges going out of a node. With respect to a dependency network, this is the number of nodes dependent on a specified node. 88

Parent node If node A of a mission-dependent network depends on node B, i.e., the operability of node B impacts the operability of node A, node A is said to be the parent node of node B. 5

Pareto-efficient A solution on a graph of the budget of an optimization versa the mission objective with the property the objective value is the best achievable for a given the budget constraint. 82, 86

Physical network A mission network that consists entirely of physical nodes. 117, 118, 122, 123, 132

Physical node A physical node would typically represent a system needed for a mission, supporting a functional task, or a waypoint an adversary must transit to reach a capability within them mission which would impact the mission's operability.. 78, 80, 86, 94, 98, 102, 104, 117–124, 126, 127

Physical node criticality A metric of extent that a physical node's operability impacts the mission's operability. 82

Prize Collecting Steiner Tree (PCST) A variation of the Steiner Tree problem in graph theory. In the PCST problem, each node in a graph has an associated prize (or reward) and each edge has a cost. The goal is to find a subtree that minimizes the total edge cost while maximizing the collected prizes, balancing between the cost of the edges included and the total prize collected from the nodes. This problem is commonly applied in the network design, such as telecommunications and transportation planning. 124

Random hierarchical network One hierarchical network for which a node at layer n is connected to a node at layer $n + 1$ or a node at layer $n - 1$ according to prespecified probabilities. 21

Red teaming A exercise where one prepares for the future by examining the options of one's adversary from their point of view in order to anticipate their likely actions. 84

Risk Achievement Worth A representation of how much the probability of critical failure increases when a particular element fails. 77

Risk Reduction Worth A representation of how much the probability of critical failure decreases when it is presumed that a particular element will not fail. 77

RiskMAP A hierarchical dependency-mapping model that identifies the cyber assets most critical to mission success. 49–53, 58, 67, 68, 163

Robust optimization A method in optimization that seeks to solve an optimization problem where there is a complication that its parameters are not known precisely. As such, one must balance achieving a desirable objective function while hedging against parameters that are known to be unknown. xxiii, 50, 135, 136, 169

Robust Optimization Solution The solution to a robust optimization problem. 169, 173, 175–177, 184

Root node The node that characterizes the operability of the entire mission. This node depends on other nodes but no node depends on it, i.e., it is not the child of any node. 118

Single child contribution One of the two qualitatively different contributions which capture the dependency of one node on other node(s). The single child contribution, captures the dependency of a single child node. For example, a single child might restrict its parent node to have a metric value no more than 20 more than the child node. The other type of contribution is a multi-child contribution. 56, 57, 59

SPLIT A piecewise-linear function. 29

Stackelberg Equilibrium The solution to a Stackelberg game with the property that the player who makes the first decision does so in anticipation of the resulting best decision of player who makes the second decision. In turn, that player makes a decision based on this. xxxvi, xxxix, 181–183, 188, 207, 210–212

Stackelberg Game A two-person game in which one player is constrained to irrevocably make the first decision. xxxix, 147, 180, 183, 202

Star and Bar Problem This problem consists of counting the number of ways one can divide n objects into k non-empty piles. This can be solved by imagining n stars arranged in a line and then putting a bar on each end and then placing $k-1$ bars between two stars so that each pile is associated with the number of stars between two bars. Since there are $k-1$ bars and $n-1$ potential locations for each, the number of combinations of doing this is $C(n-1, k-1) = (n-1)!/((k-1)!(n-k)!$. Suppose, instead, one is interested, as we are in Chapter 7, of arranging n objects into k piles where some piles might have zero objects. This is equivalent to arranging $n+k$ objects into k piles each having a nonzero number of elements and then removing one object from each pile. Hence, the number of combinations of this variation is $C(n+k-1, k-1) = (n+k-1)!/((k-1)!n!)$. 148

Steiner Prize problem A combinatorial optimization problem involving finding the shortest network of edges that connects a given set of nodes in a graph given known distances between nodes. The solution may include additional nodes, Steiner points, that are not part of the original set of nodes if doing so minimizes the total edge length. 121

Stochastic programming A class of optimization programs that involve parameters with uncertain values. 169

Strength of Dependency One of the two parameters used in Functional Dependency Network Analysis (FDNA) to capture the dependency between a node and its dependent child node. This parameter captures the extent the child node's operability impacts the parent node's operability in a mission network. This is typically abbreviated by SOD. 5, 29, 30, 51

System Identification This is the process of observing the input and output of a system and inferring the mathematical relationship between the two. 51, 60, 61, 63

System Operational Dependency Analysis (SODA) A generalization of the FDNA framework which characterizes dependency between nodes of a mission network using two parameters per link closely related to the two FDNA parameters supplemented by an additional two. 25, 49, 50, 78, 163

MIX
Papier aus verantwortungsvollen Quellen
Paper from responsible sources
FSC® C105338

If you have any concerns about our products,
you can contact us on
ProductSafety@springernature.com

In case Publisher is established outside the EU,
the EU authorized representative is:
Springer Nature Customer Service Center GmbH
Europaplatz 3, 69115 Heidelberg, Germany

Printed by Libri Plureos GmbH
in Hamburg, Germany